62 PROJECTS TO MAKE WITH a DeAD COMPUTER

and other Discarded Electronics

By RAndy SaRafan

WORKMAN PUBLISHING >> NEW YORK

Dedication

To Danica, without whose love, support, patience, and proofreading this book could not have happened.

Library of Congress Cataloging-in-Publication Data is available.

ISBN 978-0-7611-5243-9

Cover design: Robb Allen / Book design: Rae Ann Spitzenberger
Illustrations by Winnie Tom

Workman books are available at special discounts when purchased in bulk for premiums and sales promotions as well as for fund-raising or educational use. Special editions or book excerpts can also be created to specification. For details, contact the Special Sales Director at the address below.

WORKMAN PUBLISHING COMPANY, INC.
225 Varick Street
New York, NY 10014-4381
www.workman.com

Printed in the United States of America
First Printing January 2010
10 9 8 7 6 5 4 3 2 1

PHOTO CREDITS
Main Cover Photo: Jenna Bascom
Details: Sophia Su

Color Insert: Original photography copyright © Evan Sklar

Additional Photo Credits: **Jenna Bascom** p. 2, 172, 193; **Jen Browning** p. 90 (left), 95, 111, 115, 138; **Getty Images** p. 2 (color insert movie still) Silver Screen Collection/ Hulton Archive; **Evan Sklar** p. 45 (top, center, bottom right), 52, 69, 70, 72, 75, 80, 87, 89 (middle couple, bottom), 101, 104, 106, 107, 123 (center), 127, 133, 143 (second from top), 148, 162, 171, 173 (center), 181, 188, 199 (center, bottom), 203, 224, 235 (center), 239, 248, 251; **Sophia Su** p. 46, 50, 53, 56, 59, 60 (inset photo courtesy of Kathryn Millerick), 63, 65, 66, 77, 83, 86, 88, 90 (right), 92, 94, 97, 100, 102, 103, 109, 112, 116, 119, 122, 124, 129, 130, 131, 134, 144, 146, 151, 154, 156, 157, 160, 163, 166, 174, 176, 180, 183, 184, 192, 200, 205 (inset photo courtesy Melissa Scotton), 108, 209, 212, 214, 217, 220 (inset photo © Javier Montero/Fotolia), 228, 236, 241, 245.

Props and Wardrobe Stylist: Jen Everett

Hair & Makeup: Jen Browning

NOTICE TO READERS: A BRIEF WORD ON SAFETY

This book is intended to provide you with ideas and general techniques for projects to repurpose electronic equipment which might otherwise be discarded. While following along with these projects should be a rewarding and productive experience, you should always take care to use caution and sound judgment. If you are uncertain about your ability to safely perform any of the techniques described in this book, you should seek professional assistance.

Make sure to read the section on general safety (starting on page 40) before proceeding with any of the projects described in this book. Don't forget to read all of the "Safety First" textboxes you come across while reading this book. You also should familiarize yourself with the safety warnings and instructions for any tools, equipment, and other materials you may use in any of your projects. Finally, we recommend that you check with your local building and zoning authorities for any necessary building permits, regulations, codes, or other laws pertaining to your projects.

The publisher and the author of this book do not and cannot assume any responsibility for property damage or bodily injury caused to you or others as a result of any misinterpretation or misapplication of the information and instructions provided in this book, which are not meant to serve as formal interpretations of the National Electric Code®.

contents

CHAPTER 1: Tools, Techniques, and Safety

Find out what lurks in the depths of your computer, learn how to get at it, and how to dig in safely.

CHAPTER 2: Projects for a Postconsumer Dwelling

Get smart with 14 projects to deck out your desk and office, live large in your living room, and geek out your kitchen with your fried electronics.

CHAPTER 3: Fashionable Technology

Go geek chic with 12 projects to accessorize your look—because there's so much more to style than a well-placed pocket protector.

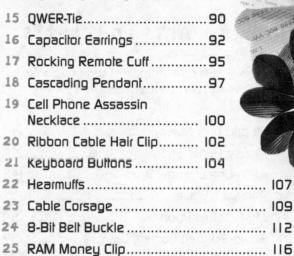

cable corsage, p. 109

Drive Bookends, p. 75

CHAPTER 4:
Fun-Fun-Fun-Fun-Fun-Fun!

My First Squiggle Bot, p.131

Robots and Aliens and Lasers—oh, my. Here are 6 clever projects you can make for yourself or a friend—just for the fun of it.

CHAPTER 5: Arts and crafts

Indulge your creative side with 9 projects that turn computer scraps into sculpture (and notebooks and wallets and pencil sharpeners and picture frames and . . .).

CHAPTER 6: Making Noise

Let's get loud: From instruments to speakers to amps, here are 6 projects to help you turn up the volume.

CHAPTER 7:
Gadget Goodness

IR camera, p. 200

From the practical to the prank-worthy, here are 10 projects to unleash the gadget god within.

CHAPTER 8: Playful Pet Projects

Feed the birds, raise an ant farm, wrangle the lizards, introduce your cat to a different sort of dead mouse and more: 5 projects for your growing menagerie.

Acknowledgments

Mouseracas, p. 181

introduction

For a large part of human history, a direct correspondence existed between the person who made a tool and the person who used that tool. This theoretical person understood not only all of the uses for this tool but also all aspects of the raw materials used to create it. He invested time and energy into this tool to aide himself in survival, and did the best he could to let as little resources be wasted.

For instance, if an early human was to expend the energy to chop down a tree, he might then use the bark to weave rope and the trunk to make a canoe. The branches may have been used for firewood, ax handles, or possibly spears. With the spears, he could then hunt animals, which in turn may have provided him with meat for food, oil for cooking, a pelt for clothing, and bones out of which more tools could be made.

There is a point to this. I promise.

When, as a modern human, we kill a computer, we simply discard it, leaving it to rot in spite of the fact that the components inside can still be used for countless other applications. Not having made these tools on which we are reliant, we don't really understand where they have come from or how they actually function.

We don't know capacitors, for instance, are largely made up of coltan (otherwise known as colombo-tantalite), a metallic ore most abundantly found deep within the Congo. Nor do we really know how the coltan is then assembled into a capacitor. And, for that matter, most of us don't even know what a capacitor is, even though there is most likely one inside every electronic device we own.

Without fully understanding these complex beings that have been created to exist among us, of course we are hesitant to "hack open" our dead computer and use its "guts" to make new and better tools. To be truthful, if given a handful of electronic components, most modern humans wouldn't have the foggiest idea what to do with them. So, not seeing any potential further value to its parts, when something breaks, we throw it away, ending its life cycle.

Unfortunately, very few objects have the luxury of leaving Earth's atmosphere. When we throw away a dead computer, we are not sending it off to some other planet, but rather we are sending it to some large hole in the ground a couple of miles away where it can stagnate in an environmentally unfriendly manner. Unlike poor Fluffy (R.I.P.), who is buried in a small hole in the backyard, your dead computer won't biodegrade anytime soon.

The author at work: Turning Tech Trash into Tech Treasure!

So if left with the choice of a) having a big toxic hole in the ground of every city on the planet filled with computers or b) converting these computers into clothing, furniture, planters, pet beds, or even just toys, I think the decision is quite simple. Our collective future is dependant upon repurposing old computers (and other "dead" technologies) to extend their life cycle and make the old new again. It is up to us to be good shepherds of the Earth through creative reuse.

Perhaps, what I mean can be best illustrated with this simple story. As a child, one of my fondest memories was discovering my parents' old record collection and learning how to listen to these strange ancient discs. From this collection of old recordings I was able to deduce that, in a distant time and space, my parents were once human. This enabled me to better understand who my parents were as people and also learn a little about myself and where I came from. Fortunately, this particular technology continued to work long enough to be able to be passed on.

With the lifespan of current audio technology being around 4 to 5 years, it makes me wonder what my children will have the chance to discover about me? Where will they discover my "record collection"? Perhaps they'll find some MP3 playlists or photos backed up on an old hard drive, but even that looks unlikely at the rate at which digital storage technology advances. If my future children learn something about me as a person and my past life, it won't be through the collection of playlists stored on an MP3 player, but what I do with the device itself after its life runs out. By repurposing it to make it uniquely my own, not only am I doing the Earth a favor, but I am reclaiming that dead MP3 player and imbuing it with a new meaning. It is no longer a poorly built consumer electronic device at the end of its brief life, but a unique and special handmade object that can be used and cherished for many years to come.

Are you now psyched to get going ripping apart old computers but still mildly intimidated and unsure where to begin? Don't worry. Start with Chapter 1, a massive crash course on the computer and electronics basics. Then dive headlong into the tools, techniques, and safety considerations you need to have and understand in order to make the various projects. From there, dig in. It's project time.

To help determine which project is right for you, each project starts with a listing of the main ingredient ("Tech Trash"), a list of supplies and an overall level of difficulty rating. This rating combines a measure of skill level, the time you need to complete the project (more involved projects get a higher rating), and safety considerations. For instance, a project rated 1 is quick, easy, and fun for any beginner, whereas a project rated 5 is an expert project which will require lots of time, skill, and a higher level of safety precautions.

To guide you along the way, Mr. Resistor Man offers up fun and useful facts relating to the project you are working on. You should also pay close attention to the "Safety First" boxes which provide important safety considerations relating to the project at hand.

Mr. Resistor Man Says: If you like the look of me—what's not to like?—you can make your very own Mr. Resistor Man pocket pal on page 129.

CHAPTER 1

Tools, Techniques, and SAFETY

> **Find out what lurks in the depths of your computer, learn how to get at it, and how to dig in safely.**

L et's get started. First, look to the right, and start by familiarizing yourself with the anatomy of a dead computer. It's like you're in medical school, studying a cadaver. You should get to know the parts, how they're connected, and how you can safely detach certain "ligaments" and other key components. But in order to really be a good study, it'll help to have an actual dead computer to poke around in.

"SHOPPING" FOR A DEAD COMPUTER

S ometimes the farthest you have to go to find dead electronics is your own garage, attic, or storage closet. But if you find a project in the following pages for which you are lacking key parts, don't worry. Just pick up the phone, call everyone you know and start asking if they have what you are looking for— and you might explain why you want it, too. You'll be surprised and amazed at what people are hoarding in their houses and even more surprised and amazed at how eager people will be to unload their broken electronics on you if

they think you'll be putting it to good use. I've gotten a lot of wonderful junk through this method.

If it turns out that asking friends and family for their e-waste is a fruitless endeavor, look online (yes, consult your live electronics to locate the dead ones). One great website to check out is Freecycle.org whose mission it is to regionally connect people who want free stuff with people who want to give away stuff (a win-win situation). Simply post a message to your local Freecycle list saying that you are looking for broken computers (or NES systems, or . . .), go for a long walk, and return

THE ANATOMY OF A COMPUTER

A computer is any device that computes. In the case of this book, it is an electronic device that is programmable and can store and process information (in plain English, it's anything that you can check e-mail and play video games on). Here's a cross-section so you can see what's what.

POWER SUPPLY

The power supply converts wall current into many different lower-voltage sources to provide the correct voltage to a number of computer components with different power requirements. Inside is an array of parts like capacitors and toroidal coils. Read the safety section on handling high-voltage capacitors before you open the power supply.

THE MOTHER BOARD

The largest circuit board inside of a computer, the mother board houses the "brain" of the computer—the microprocessor—which does all of the computation based on the information it gets from a wide array of components like the hard drive and RAM card.

COMPUTER EXPANSION CARDS

These cards plug into your computer and can be accessed through ports in the back of the case and add functionality like sound, video processing, and Internet. They often contain a number of special parts such as audio transformers and computer chips that you wouldn't find elsewhere.

COMPUTER FAN

Fans inside of a computer are used for cooling. They are powered by a motor and typically spin in only one direction.

OPTICAL MEDIA DRIVE

The two most common drive types are CD-ROM and DVD. A drive is used to read compact discs which can store data, music, or movies. You can find motors, gears, and circuit boards inside.

FLOPPY DISK DRIVE

This drive stores small amounts of data using magnetic recording onto portable "floppy" storage disks. It's a good source for an array of small motors, gears, and components.

HARD DISK DRIVE

The hard drive is the permanent, long-term storage (memory) where you save all of the files on your computer. It has actuator arms, magnets, circuit boards, and shiny "platter" discs inside.

RIBBON CABLES

These are flat and wide cables used inside of your computer to connect together the different components inside of the computer.

RAM CARD

A RAM card is a single-circuit board that temporarily stores in memory the important information the computer needs to run. In contrast to the hard disk drive, it records "short-term memory."

to be inundated by messages from people who are ecstatic to unload their junk on you. The "free" section on Craigslist.org is another great resource to explore, too. Check back from time to time to see who's moving and what they're getting rid of along the way.

Another option, for the more adventurous: walk or bicycle around your neighborhood on garbage day and see what people dump out on the curb (especially on big item collection day). You will be amazed at the amount of electronics that get dumped next to the trash bins once you start paying attention. Also keep an eye out for garage and estate sales. While buying things at these events may cost you a few dollars, it is still a cheap and reliable treasure trove of obsolete electronics.

Ultimately, if you are dead set on getting your broken electronics for free, the last thing you can try is asking for handouts at your local recycling center. Some of these centers are more open to give people broken electronics

than others, but I have had luck when I tell them that I am an artist and it was imperative that I get broken electronics in order to make my art. Everyone loves supporting the arts!

However they turn up, once you get the broken electronic item of your dreams, start to get to know it. Respect its power (even in its deadness) and know your way around its circuitry.

The Basics

There is no way that I can teach you all, or even a fraction, of the electronics necessary to fully understand how a circuit functions. However, I *can* give you a crash course to help you start to formulate an idea of what is inside those dead computers of yours—and how you can use this new-found knowledge for good.

CIRCUITS AND CURRENT

A circuit is a complete path formed by various conductors and components through which electricity can pass.

To better grasp what this means, we need to understand the two types of electric current that we will encounter: AC (alternating current) and DC (direct

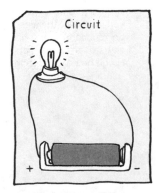
Circuit

Not a Circuit

current). The difference between the two is that with DC power the electrons always flow in a constant direction from the source of power to ground. However, with AC power, the electrons are constantly and rapidly alternating direction through the circuit and at any given time can be traveling either way through the circuit.

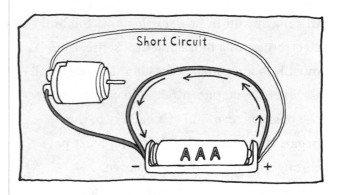

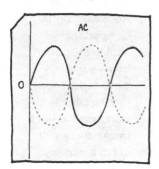 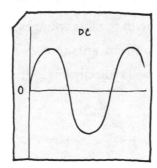

AC Power

Current that comes out of a wall is AC power (as AC is better suited for traveling long distances through a wire) and current that comes out of a battery is DC. Throughout the course of this book we will be dealing primarily with DC power, as most electronics mainly use this type of power.

DC Power

There are a few basic things we need to understand about DC electricity before we can start controlling its flow with various electronic components. The first rule we need to keep in mind is that *electricity always*

follows the path of least resistance to ground. What this means is that a source of electricity is "looking" for the easiest way to discharge all of its power. For instance, if a source of electricity has the choice of flowing through a motor or a bare wire to get to ground, it will flow through the wire because that provides the least resistance. At first glance, this may seem perfectly reasonable, because we are helping the current get to ground faster, but this is *bad.* Not only will the motor not spin because electricity is not passing through it, but we have just connected power directly to ground and shorted the circuit! Go get your fire extinguisher!

The circuit shorted because all of the energy reached ground without being used. This brings me to the next point that cannot be stressed enough: *In a circuit, all electrical energy must be used!*

If electricity passes through the circuit without being used by anything, a much larger than expected current reaches ground, and this may result in fires, explosions, and broken electronics. In a more practical sense, there must be something in the circuit to use up all the energy you put into it. You cannot connect the positive terminal of a battery to the negative terminal. The electricity must pass through components and be used.

The components in your circuit will try their best to convert any extra energy to heat. However, if you put far more energy into your circuit than your components can ever use, they will have to work so hard in converting this energy to heat that they will burn out and stop working. If this happens, you should congratulate yourself for having "fried" your electronics. Fried electronic components are often the cause when computers stop working.

To keep your components from getting fried, it's important to add other components to your circuit that add resistance and are properly rated for the amount of electricity you are putting in. That means that both the voltage and current rating on the component you are using need to be rated within range of the amount of electricity you are providing. Usually, providing enough resistance is not a problem when working with the low voltages and currents that we will typically be using.

TINY COMPUTER PARTS

It would be irresponsible of me to continue talking about electronic components if you haven't even met them yet. So, without further ado, let me formally introduce you to the little parts that help control the flow of electricity through a circuit.

wire

Wires are the metal threads upon which electricity flows. The wire you will using in this book when working with electronics is insulated, which means that it has a plastic jacketing over its metal conductive core. This allows many

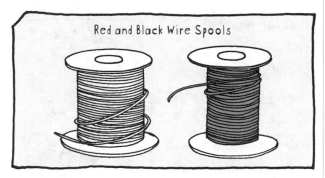

Red and Black Wire Spools

wires to get close and personal with one another without crossing and causing a short circuit.

You will typically work with two colors of wire, red and black. Red is traditionally used to represent positive and black is used to represent ground in the circuit. Though there is little real difference between the red and black wires structurally, it is good form to stick with red as positive and black as ground or "negative." This will help in troubleshooting (and allow other people to easily lend you a hand).

Aside from color, the other important thing to keep in mind when selecting wire is the type of wire it is. The two most common types are solid and stranded wire. The difference between the two is that solid wire has a single solid core of metal inside of the jacketing and

stranded is made up of lots of little metal fibers woven together.

You will mainly use solid wire when building circuits because it is easier to solder—it doesn't fray at the ends and can be bent into a fixed shape. This makes it ideal for circuit building, hacking, and a number of craft applications. The downside of this type of wire is that it can get brittle and will snap if it is bent too much.

The benefit of stranded wire is that it is flexible and can be bent around without getting brittle and snapping. It is for that reason that power plugs, computer cables, and headphones (where wires are constantly being bent around) use stranded wire. In this book you will use stranded wire when making things that are going to be bent around or squished.

Switches

A switch is a mechanically controllable break in a circuit. Think of it this way: When you flip a switch in one direction, there is a break in the wire preventing electrons from flowing through the circuit and when you

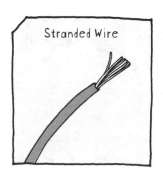

Stranded Wire

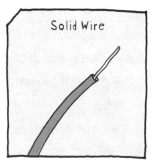

Solid Wire

SWITCHED ON, BABY

Switches typically add little or no resistance to a circuit and should not be used to connect the positive and negative terminals of a battery directly together.

flip the switch back in the other direction, you reconnect the wire and allow electricity to flow again. A switch, therefore, is simply a mechanical device that opens (cuts) or closes (mends) a circuit connection.

There are a few types of switches you may encounter while working with broken electronics. One of the most basic is the slide switch. We have all seen these switches before. These generally have a little plastic box that you can slide across a track to turn something on or off. These are commonly used as power switches.

We also commonly encounter push-button switches. These types of switches are triggered by pressing in a button. There are two common types of push-button switches: momentary and toggle. Momentary only works so long as you are pressing down the button, and the moment you stop

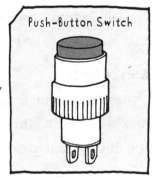

Push-Button Switch

pressing down, the momentary switch stops working. A toggle switch, on the other hand, switches between on and off with every button press. Once you press the button (like your TV power switch), it will remain on until you press again to turn it off (and vice versa).

Push buttons aren't the only type of switch

Toggle Switch

that can be toggled on and off. As the name would strongly imply, the toggle switch can be, too. The most common toggle switch is the ubiquitous wall-mounted light switch. Quite simply, if you "flick" or toggle the switch up, power is connected and the lights turn on and if you flick it down, the lights go off (this may be opposite in your household).

Now that we know how to turn the circuit on and off, it's time to learn what components make up the circuit.

Resistors

One of the most common components found on a circuit board is a resistor. Resistors are small pill-shaped tubes with four colored bands wrapped around their casing. They are responsible for "resisting" or restricting the flow of electrical current within a circuit by converting electrical energy to heat. Resistors have different values that can be deciphered based on the pattern of color bands wrapped around it (see chart). As the resistance value goes higher, the current in the circuit decreases.

Resistors are *nonpolarized*, which means that in a DC circuit (in which electricity is only

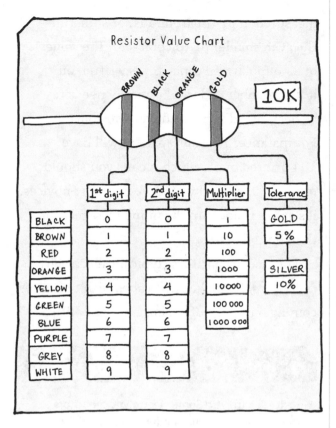

Resistor Value Chart

BROWN BLACK ORANGE GOLD

10K

	1st digit	2nd digit	Multiplier	Tolerance
BLACK	0	0	1	GOLD 5%
BROWN	1	1	10	
RED	2	2	100	
ORANGE	3	3	1000	SILVER 10%
YELLOW	4	4	10000	
GREEN	5	5	100 000	
BLUE	6	6	1 000 000	
PURPLE	7	7		
GREY	8	8		
WHITE	9	9		

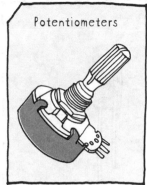

Potentiometers

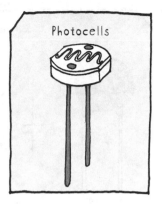

Photocells

flowing in one direction), the resistor can be placed in the circuit forward or backward and allow electricity to pass through in the right direction.

A special type of resistor is a variable resistor, which, as its name suggests, has resistance that can vary. A common type of variable resistor is called the potentiometer (or "pot"), which looks like a round disc with a twisty knob coming out the center. These are used in electronics as stereo volume knobs and just about any other "twisty knob" you may encounter. By turning it you are changing the resistance in the potentiometer from 0 ohms resistance to the maximum value written on the casing. So, for instance, if it has 10K printed on the back, by turning the dial you would be able to change the resistance from 0 ohms all the way up to 10K ohms (10 kilo ohms or 10,000 ohms).

Aside from potentiometers, a number of other components, especially certain sensors, change the resistance in a circuit. One ubiquitous example of a sensor that changes resistance is a photocell (or "LSR," which stands for light-sensitive resistor). The amount of light that shines on the surface of the photocell affects how much resistance it has in the circuit. In fact, the more light that shines on it, the lower the resistance is.

capacitors

Another commonly found component is a capacitor. There are a number of different types of capacitors, but they all basically function the same. A capacitor stores a set

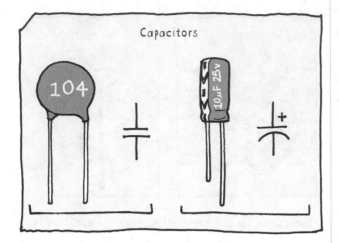

Capacitors

104

10μF 25v

amount of charge until it is completely full, and when the current in the circuit drops, the capacitor releases (discharges) the stored charge back into the circuit. Imagine a water storage tank that fills up when there is excess water available; when there is a shortage, it then dumps out (discharges) all of its excess reserves back into the water system. This is how a capacitor operates.

The two types of capacitors you may encounter are polarized and nonpolarized. A polarized capacitor will only let electricity flow through in one direction and can only be put into the circuit a certain way (one pin going to ground and one to the positive source). A nonpolarized capacitor can go into the circuit either way.

Nonpolarized capacitors are typically ceramic disc capacitors and look like small, round clay discs with two wires coming out. This type of capacitor is normally

measured in μF (microfarads) or sometimes the even smaller pF (picofarads). The value of a nonpolarized capacitor is written on it in a number code and typically needs to be deciphered using a conversion chart. For instance, a 0.1μF capacitor will have 104 printed on it, which means you should multiply 0.00001 by 10,000. For most novices, it's easier to look up the number correlation than do the math.

Polarized capacitors (typically electrolytic capacitors) look like large tubes with two wires coming out. They often have a stripe with a

⚡ SAFETY FIRST

>> It's Electrolytic! Some electrolytic capacitors can store a very large, painful, and potentially dangerous amount of electricity. You should never handle these capacitors without discharging them. Fortunately, the voltage rating is printed on all electrolytic capacitors to help you determine whether or not you need to discharge it.

Sometimes a capacitor will have an "F" rating printed upon it (as opposed to μF). This stands for farad. A farad is a very large electrical charge. In fact, a farad (F) is one million times greater than a microfarad (μF). Capacitors measured in farads are often called "super caps" because of the amount of charge they store. Super caps are not encountered very often (I once found one in a particular model of PalmPilot), but if you come across one, treat it as though it were a high-voltage capacitor and discharge it before handling.

For more information on discharging a capacitor, refer to the Safety section, page 40.

DISCHARGING A CAPACITOR

Circuit boards can be dangerous, the primary reason being that they house very powerful capacitors. Here is an easy, though non-standard, way of discharging a capacitor. *Note:* Do not use this method when disassembling a CRT monitor.

1. CUT AND STRIP

Take a strand of Christmas lights and cut away one bulb so there are two 5" insulated wires running away from the light. Use wire strippers to strip the insulation away ½" from the end of each wire.

2. PROTECT YOURSELF

Put on rubber dish gloves for added protection.

3. LOCATE THE TERMINALS

Being careful not to touch any exposed metal on the underside of the circuit board, locate the terminals on the underside of the circuit board to which the large capacitor is soldered. You can generally tell which ones these are because they will appear to be directly below the capacitor and there won't be any other solder points relatively nearby.

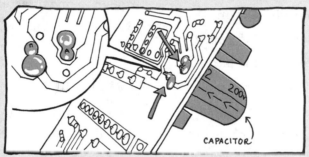

CAPACITOR

4. DISCHARGE

Hold the Christmas light wires between your gloved fingers on the insulated parts (do not touch the exposed wire) and place an end of the exposed wire on each of the capacitor terminals. If the capacitor is full, the light will quickly flash on and off, indicating that it is now mostly drained.

Note: If the light does not turn on when you touch the wires to the terminals, touch the exposed wire to a few different nearby solder terminals for safety's sake to make certain you are discharging the right connections. If after a dozen or so different attempts, the capacitor appears not to be charged, it is always good measure to "discharge" it anyway by bridging the terminals with a long screwdriver.

5. DRAIN

Just because the light went off does not mean that the capacitor is completely discharged. Drain the small amount of charge that is left by bridging the capacitor terminals with the tip of a long screwdriver. Do not touch the metal part of the screwdriver.

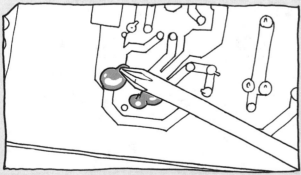

6. WAIT, REPEAT

Sometimes the capacitor can refill with small amounts of electricity hidden in other parts of the circuit. Wait three minutes and repeat Steps 4 and 5.

minus sign printed on one side to indicate which wire connects to ground, and show their voltage and capacitance rating. The capacitance rating is almost always in μF.

Diodes

Aside from the electrolytic capacitor, if we had to choose one quintessential polarized component, it would be the diode. A diode is a small, thin cylinder with two metal leads sticking out of each end and a band around one end to indicate the negative lead (the cathode). Electricity can flow in only one direction through the diode, from the positive end (the anode) to the negative (the cathode). In other words, if you were to insert a diode backward into a circuit, no electricity would pass through the diode, and the circuit would stop working properly. If you were then to correct this and flip it the other way, the electricity would flow freely through it from

positive to ground. Allowing electrons to pass through in one direction and not back in the other makes diodes ideal for a number of applications, such as protecting circuits from harmful voltage spikes.

People often think of LEDs as tiny lightbulbs, but they are actually a special form of diode that emits light (photons). Hence, it is not surprising that LED stands for "light emitting diode." Because LEDs are diodes, they are polarized and electricity can only flow through in one direction. Typically the longer leg connects to the positive supply and the shorter to ground.

Transistors

The invention of the transistor in 1947 sparked the dawn of the computer age. A single transistor could do the amount of computing that it used to take an array of more than 40 air-cooled vacuum tubes to do.

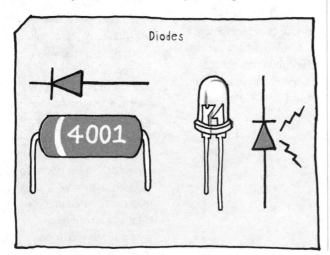

Diodes

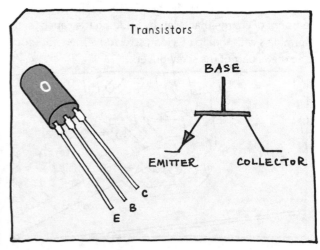

Transistors

Packaging Is Everything
It is important to keep the packaging your transistor comes in so that you know how to properly hook it up. If you accidentally tossed it already, look up data sheets online, because the pins for the base, emitter, and collector differ from transistor to transistor.

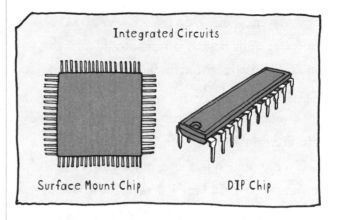

Integrated Circuits

Surface Mount Chip   DIP Chip

In other words, transistors made digital logic cheap, easy, and compact. Transistors are little devices that generally look like half cylinders with three legs poking out the bottom.

The easiest way to think of a transistor is to envision it as a water valve controlling a garden hose. When you apply a positive voltage to the base pin, it is as though the valve is turned all the way on and water can flow freely through the hose, just as electricity is allowed to flow through the collector and emitter pins. When voltage is removed, the valve is shut and the water stops, or, in actuality, the flow of electricity is cut off. If only half the electricity is supplied (say, 2.5V from a 5V power source), the valve only allows half the amount of water to flow, and it would follow that only half the electricity available will be allowed to pass through the transistor.

> **Mr. Resistor Man Says:**
> By combining a photocell (the light-sensitive variable resistor mentioned earlier) and a transistor, we will be able to make a switch that turns on LEDs based on how dark it is in the room. This circuit is later used in the Postindustrial Night-Light, page 228.

Integrated Circuits

If we say that the transistors triggered the computer age, then it would not be a stretch to say that the integrated circuit made the personal computing revolution possible. You may have heard of integrated circuits referred to as semiconductor chips, ICs, or just "chips" for short.

An integrated circuit is essentially an enormous number of minuscule transistor circuits arranged onto a small wafer of silicon. The beauty of this is that transistors can be arranged in such a vast combination of logic circuits that their possibilities are only limited by the number of transistors you can fit inside a chip. This abundance of possibility has resulted in a multitude of chips all suited to an array of different tasks.

For instance, one chip might have an MP3 decoder built inside, whereas an identical-looking chip might actually be the driver for a digital alarm clock.

There are two common types of integrated circuits that you may encounter. The first is a DIP-chip, which is a long box with two rows of metal pins that stick down. These are often pressed into sockets (that they can be easily removed from) or soldered to the underside of the circuit board. This type of connection is called through-hole mounting because the chip passes through the board and is mounted on the other side.

The other type of IC is a surface-mount chip. These are small chips with tiny legs soldered onto the surface of the circuit board. These "legs" (pins) are very important. By making an electrical connection with the legs, you can access the circuit hidden inside. Each pin is electrically connected to a different part of the circuit built inside the chip.

ELECTRICITY

In all of this talk about circuits, electricity, and electronic components, we left out the most important part and the one that tends to truly stump most beginners—that is, getting electricity into the circuit.

Power Supply

The power supply applied to a computer comes directly from the wall socket. This means that the amount of electricity flowing into a computer is 120VAC. This is a very high voltage and potentially extremely dangerous. Fortunately, the wall current goes directly into the computer's ATX power supply, where it is converted to less-dangerous low-voltage DC power supplies that range from 3.3VDC to 12VDC.

A computer cannot run directly off the electricity coming from a wall socket. The electricity needs to be transformed to a lower voltage and current. This is the same for all devices that run off DC power, such as printers, monitors, and scanners. Although they appear to be running directly off wall current, the wall current is almost immediately converted by the device to a low-voltage DC supply.

So, if you want to power up the low-voltage circuits that you build, DIY-style, you should never plug them directly into the wall. This will typically blow fuses, start fires, and/or possibly electrocute you.

Transformers

To power low-voltage devices using wall current, you will need something to convert these high AC voltages into low DC voltages. Typically what you need is a transformer. Not to be mistaken for a car that folds into a robot, this device is a boxlike plug that inserts directly into a wall socket. This is sometimes referred to as an AC adapter and "wall wart," because it looks like a large, ugly growth coming out of the wall.

Transformer

Transformers almost always are labeled with their initial power rating, which for those of us in North America should be 120VAC, and their secondary power rating, which for most consumer electronics tends to be between 3VDC and 15VDC, with a current rating of anywhere between 100mA to 3 amps. Our ideal power supply for working with electronics is 5VDC at 500mA, which is by coincidence currently the standard power supply for most cell phones. We will also be using 9VDC and 12VDC supplies in this book. Try to avoid using anything with a current rating of over 800mA, because at that rating, the current starts to get too high to handle safely.

Despite the fact that it is plugged into the wall, the exposed wires coming out of a transformer are usually at a safe voltage for you to handle. For instance, you can clip the adapter from the charger that plugs into your cell phone, strip back that wire, plug it into the wall, and handle it without danger (assuming the secondary power rating is within the ranges discussed in the prior paragraph). This may take some getting used to, as we are trained not to touch any bare wires that are plugged directly into the wall.

However, to reiterate, if the secondary voltage is about 15V and/or the secondary current is above 800mA, don't handle these wires (or use the transformer in the first place). And you should also avoid using any transformer or plug that does not have a power rating written on it or that has a rating you do not understand. It is always better to be confused and safe than inquisitive and sorry.

BATTERIES

Moving away from the wall, we arrive at power supplies that are portable. Of course, I refer to batteries. Chances are, if you are like me, you not only know what a battery is but you are constantly scouring the house for them. We all have batteries lying around, but they always seem to be in the last place we look. For me, that tends to be in the remote control for the TV (sorry, honey!).

coin cell Battery

The smallest battery you will encounter is the coin cell battery. These come in an array of sizes and usually carry a voltage of 1.5V or 3V (this rating is typically written on the battery). A standard coin battery is the 2032, which is a 3V battery located on almost every PC motherboard to keep the CMOS (the part of the computer that stores clock information and system settings) powered when the computer is turned off.

To use coin batteries, you can either use a battery holder (like the one on the motherboard) or tape wires directly on the surface of the battery. When attaching wires to

Coin Cell Battery

3V
DL2032
+

Coin Battery Holder

⚡ SAFETY FIRST

>> **Never solder directly to a battery! It will explode and send superheated corrosive fluid flying everywhere. Instead, get battery holders (they're available for just about any battery type) that you can safely solder. Remember to solder the holder to your circuit before you put the battery into it.**

---

coin batteries you should be careful, as the "+" side extends around the side edge of the battery.

cell-Type Battery

One of the most common power supplies you will use to work with electronics in this book is the standard cell-type battery. The important thing to remember about cell-type batteries ranging from AAA to D is that despite their different sizes, they are all 1.5V. The different sizes relate to the amount of power they can store. This means the larger the battery, the more power it will hold. If you want something to be powered for a long time, you should use a larger battery.

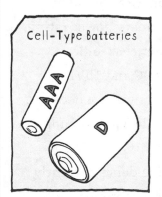

Cell-Type Batteries

AAA

D

Battery Holders

9V Battery

Last but not least, we have the wonderful 9V battery. I love 9V batteries. They are relatively small, they tend to last a long time, they produce a fair amount of current (so don't lick them!), and they can be easily attached to a circuit using a small battery clip. Most importantly, if you couple a 9V battery with a voltage regulator, you can down-convert the 9V source into a regulated 5V, 3.3V, or even 1.5V power source, which makes the 9V battery an ideal long-lasting solution in an array of different circuits.

Voltage Regulator

A voltage regulator is a device that takes in a source voltage on one side and releases a constant steady voltage on the other. Typically the source voltage has to be greater than the output voltage.

Voltage regulators look like black square boxes with three "legs" and a metal tab with

Voltage Regulator

7805

Need to Know
If you apply 12V or greater to a 7805 voltage regulator, fasten a heat sink to the metal plate in the back. A heat sink is a special metal bracket that helps dissipate heat and keeps the regulator from overheating and getting "fried" (or worse).

a hole in it. The side with the metal tab is considered the back and the one with the legs the bottom. The most common voltage regulator (and the one we will use in this book) is the 7805, which can accept between 8V and 15V and output a steady 5V of electricity.

Multimeter

The mighty multimeter measures a multitude of different things. In this book we will primarily use it to measure voltage, check polarity (which wire is positive and negative), measure the resistance of various electronic components, and check for continuity.

To measure voltage, set the dial on the multimeter to the "V" with two lines, one dashed and one solid.

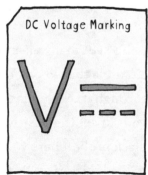

DC Voltage Marking

V

Multimeters should be able to measure a range of different voltages (very cheap ones,

Need to Know
Depending on the multimeter, you may not have the range markings. On some multimeters, you need to press the "range" or "mode" buttons to toggle the voltage measurement range.

Correct Reading

Reverse Polarity Reading

however, have only one range). When making circuits in this book, you will be measuring electricity in the 1- to 12-volt range. This range can be selected by turning the dial to the "20" marking. If you select the "2" marking, the higher-range voltage such as 9V and 12V will give an out-of-range reading.

Voltage Reading

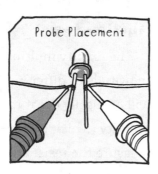

Probe Placement

To see what I mean, first make sure your circuit is turned on and electricity is flowing through it. Next, place the red probe where you expect the positive voltage to be and the black probe where you expect your ground to be. If electricity is flowing correctly, it will give you a reading telling you the amount of positive voltage flowing between those two points. If there is too much electricity flowing

for your range setting, it will tell you that you are out of range.

If you are unsure of the polarity of your circuit—in other words, if you don't know which wire is positive and which is ground— use the multimeter to figure it out. While the meter is still in DC voltage mode, simply place the red probe on any single wire and the black probe on any other wire. If you get a reading that has a "−" sign in front of it, you are reading a negative voltage. This means you have the probes backward, since you should be reading a positive voltage. Switch the probes and now look at the voltage measured. It should be the same voltage you just measured,

SAFETY FIRST

>> **Never use your multimeter to test high-voltage AC or DC (like wall current, large capacitors, or cathode ray tubes). This is very dangerous (like sticking a fork in a wall socket) and will not end well for you or your meter.**

but positive. Now it is safe to say that the wire connected to the red probe has a positive voltage supply and the black probe is ground.

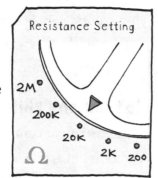

Resistance Setting

2M

200K

20K

2K 200

Ω

Another helpful thing the multimeter can do is test the resistance over an electronic component such as a resistor. This is particularly useful when you have a variable resistor like a potentiometer or photocell and need to find out its exact value. For instance, while measuring a photocell, you will see the resistance the multimeter is

Resistance of Photocell with Light

Resistance of Photocell without Light

displaying change depending on how much light is present on the photocell's surface. Knowing a part's resistance (or "range" of resistance) is handy when working with LEDs and other parts that require a certain amount of extra resistance placed in series with it. This way you can measure that extra part and be certain that you are providing

the correct amount of resistance and not going to fry the circuit.

You can easily test a component's resistance by turning the meter to the resistance-testing mode, which is the mode that has an ohm symbol (Ω) next to it. To take this measurement correctly, the resistor must not be receiving power and it can't be attached to any other parts in a circuit (it must be on its own). Start with your multimeter setting in the "20K" range. Place one probe on each lead of the resistor you are trying to test and look at the reading. If the reading is out of range (it will likely produce a "1" on the left side of the screen), try increasing or decreasing the range until you get a reading. Once you have a reading, the number it provides will correspond to the range that it is in. So, if you read 20 while in the megohm range, the resistor value is 20 mega ohms. However, if you read 20 in the ohms range, the resistor is only 20 ohms.

You can also use the multimeter to test continuity (whether or not electricity can flow freely between two points). To do this, set the

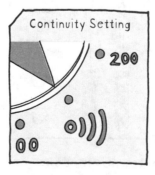

Continuity Setting

200

Continuity Testing

dial to the picture of a sound wave (or similar) and connect a probe to each of the spots that you want to check to see if electricity can freely flow between. If electricity can flow freely between these points, your meter will beep and/or change the reading on its display from 1 to 0 (some displays will say "OL," which stands for "open loop"). If electricity cannot flow freely between these points, there will be no beep or change in display.

Hand Tools

From screwdrivers to pliers to hammers to saws, here is a set of tools that bring a whole new meaning to computer hacking. Though some of the tools below are used less frequently than others, these are the ones I recommend having in a well-stocked techie toolbox.

SCREWDRIVERS

One of the most important tools at your disposal is the trusty screwdriver. It is good to have an array of different-sized and -shaped screwdrivers so that you can unfasten whatever bolt may come your way.

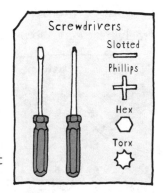

You may already have a screwdriver or two lying around somewhere. If you don't, you can find any of the screwdrivers listed here at any basic hardware store.

Slotted and Phillips Screwdriver

The two most common and frequently used types of screwdrivers are the slotted or "flathead" and the Phillips or "crosshead" screwdrivers. Though the Phillips will be called for more frequently when working with electronics, it's good to have both types in your toolbox and in an assortment of sizes.

Jeweler's Screwdriver

Another important and often used type is the jeweler's screwdriver. These are basically specialized Phillips and slotted screwdrivers capable of removing very tiny screws. You can usually buy these in sets of six or more.

Hex Wrench

After the Phillips and slotted, the next most commonly encountered driver is the hex wrench. This is used for removing bolts with hexagonally shaped sockets in the center (famously provided in almost all Ikea home-furnishing kits). These sockets come in many sizes and, as a result, hex wrenches are usually sold in sets of graduated sizes.

Torx Wrench

The last type of frequently encountered screwdriver is the Torx wrench. The Torx wrench is designed to remove, as one would guess, Torx bolts. These are star-shaped bolts commonly found inside printers, cell phones, and other modern devices that the manufacturer does not want you to easily take apart. It is essential to have a Torx wrench to accomplish easy disassembly. Try to find one with an assortment of changeable bits for maximum usability.

PLIERS AND CLAMPS

Fortunately, in this book, you'll do more than twist things (you'll grab, grip, and pull things, too!)—and your toolbox should reflect that.

Pliers and Clamps

Flat-nose Pliers | Needle-nose Pliers | Vise Grips | C-clamp | Table Vise

Flat-Nose Pliers

The most common tool used for grabbing things is the flat-nose pliers. This is an ideal general-purpose type of pliers, lightweight and good for many functions. For instance, you may use it to snap plastic pieces off the inside

casing of your computer mouse or to steady a bolt while you unscrew the corresponding nut.

Needle-Nose Pliers

Several projects call for needle-nose pliers—they're exactly like flat-nose pliers, but skinnier, and with longer "noses." Thanks to their shape, needle-nose pliers are great for both grabbing components in tight spaces and gripping very small parts.

Jewelry Pliers

Jewelry pliers are also used on occasion. They are exactly like flat-nose pliers, except they don't have the serrations on their pinching surfaces. This allows you to grab and manipulate jewelry wire (or other delicate materials) without scratching it or leaving marks (which, as you might imagine, is helpful when you are considering the aesthetics of what you're making).

Vise Grips

If you really want to make certain that you have a solid hold of something, use locking pliers, or "vise grips." This type of pliers has an adjustable locking mechanism that allows it to strongly clamp onto whatever it grips. This will come in handy when tapping the threaded rod through the hole in the phone handset while you're building the Phone Rack. Twisting it in with your hands may prove to be a lot of effort, as the rod may keep slipping in your grip. So, to save yourself hard work, clamp the pliers onto the rod and use them as a nice lever to twist the rod into place.

C-clamps and Table Vise

Last, if you need to hold something still and keep both hands free to work, there are two options: You can rig up a couple of C-clamps, which are ideal for holding flat surfaces down on a workbench, or use a table vise, which sits atop the workbench and can hold many different objects in place.

HAMMER, CHISEL, AND HACKSAW

Okay, so beyond twisting and gripping, let's get to the good stuff. Sometimes, in order to get at a computer component, you just need to break, bend, or smash something in the way.

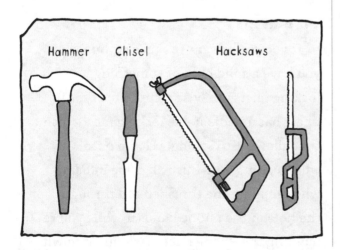

It's advisable to keep around a few handy tools for doing this.

Hammer

The ultimate tool for breaking, bending, and smashing is the claw hammer (this is your standard hammer with the two prongs on the back for nail removal and other prying). Chances are you've seen one of these before and roughly understand the principle of its operation.

Chisel

A good friend of the hammer is the chisel. The fundamental operating principle of the chisel is to position the chisel's point on whatever you are trying to smash. Once situated, use your hammer to hit squarely upon the chisel's head (the nonpointy end) while avoiding your fingers. Easy as that!

Hacksaw

The right hand of any hacker wouldn't be complete without a hacksaw. So if the hammer and chisel don't cut it for what you are trying to do (like get the darned case off that printer), you can rest assured that a hacksaw will. It can cut through almost anything, but is especially ideal for cutting things like metals. In a pinch, it will cut through plastic like a warm knife through butter. For smaller operations, you can get a handheld hacksaw with a short blade protruding from the handle (I highly recommend this).

SCISSORS, CRAFT KNIFE, AND RAZOR BLADE

A hacksaw isn't always the most precise cutting tool. For more precise application (like cutting fine curves in paper or slits in the plastic circuit board inside a keyboard), you should consider adding other cutting tools, such as scissors, craft knives, and razor blades.

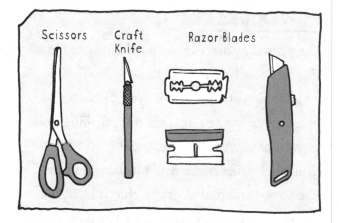

Scissors

If you do not know what scissors are used for, please put down this book and go back to watching TV. That said, make sure the scissors you have at home have nice sharp blades. If you plan on making any projects in this book that involve fabric (like the Squid-Skinned Case or Alien Appreciation Key Chain), consider investing in sewing scissors, which make it easier to cut neatly through fabric.

Craft Knife

One of the most useful cutting tools at your disposal is a craft knife. Craft knives tend to have a penlike shaft and interchangeable blades. The standard blade is wedge-shaped and razor-sharp on one side—great for cutting paper, mat board, photographs, and anything that may require a clean, precise cut.

Razor Blade

Another cutting tool you should have carefully stored away is a razor blade. These are good for making clean cuts through light material and carefully splitting seam lines in plastic casings. You should always be very careful handling these, as they are sharp and

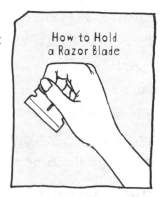

can easily cut you—your standard razor blade has no handle and is gripped firmly between your thumb and the side of your pointer finger.

CUTTING PLIERS

Last, but definitely not least, there are two hand tools that have mild identity crises. The diagonal cutting pliers and wire strippers are somewhere between pliers and scissors and they are *indispensable* in repurposing old electronics.

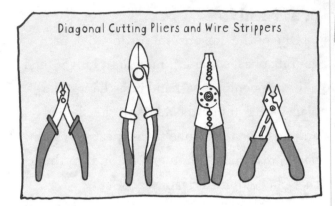

Diagonal Cutting Pliers and Wire Strippers

Diagonal Cutters

I cannot emphasize enough the importance of having a pair (or two) of diagonal cutting pliers (or "diagonal cutters") in your toolbox. They can be used to cut wires, plastic, thin metal rods, and a host of other things. They can also be used to strip the jacketing off wires. When building small objects, taking apart circuit boards, and ripping open plastic casing, the diagonal cutting pliers are a godsend.

Wire Strippers

Slightly less versatile but equally important are wire strippers. This tool is similar to diagonal cutters, except it has a series of grooves for different gauges of wire so that you can strip the jacketing without cutting the metal part of the wire. Simply drop the wire into the correct groove for your gauge of wire (gauge is a wire's thickness), close the pliers, and give the wire a tug. The jacketing will be stripped right off. They also have blades that can be used to completely cut through wire if needed.

Power Tools

Sometimes human power just won't cut it and you'll need a tool or two with more oomph. Though you won't be using very many power tools in this dead electronics journey, there are a few important ones to familiarize yourself with.

SAFETY FIRST

>> Always wear eye protection when using power tools.

POWER DRILL

The primary power tool that will be used in this book is the power drill. A power drill is a handheld rotary drill that can be battery-operated or corded. It is used for drilling holes of various sizes into different surfaces. To drill a hole, insert the drill bit that you would like to use and make certain that it is locked correctly in place by tightening the chuck (the clamping teeth that grip the bit in place). Set the drill to spin forward (clockwise), and ease your finger onto the trigger until the bit starts

Power Drill and Drill Bits

to spin at a medium speed. Slowly lower the point of the drill bit into the object you are drilling and continue until the bit has passed far enough through to make the proper-sized hole. Without slowing down the rotation of the drill, back the bit completely out of the hole, turn the drill off by releasing your finger, and admire the handiwork.

SAFETY FIRST

>> **Get a grip: Always firmly clamp the object you are drilling to the work surface using C-clamps or a table vise.**

Drilling different-sized holes using a power drill requires appropriate-sized drill bits suited for the material you are drilling. For instance, when drilling through wood, you would want a drill bit designed for wood. Your local hardware store should be able to help you determine what is right for your project.

HEAT GUN

Aside from a power drill, we're also going to use a heat gun. A heat gun is essentially a supercharged hair dryer capable of producing temperatures comparable to those of a kitchen

Heat Gun

SAFETY FIRST

>> **Heat Gun 101: Always remember to point the heat gun away from yourself—it gets extremely hot very quickly. Also, keep in mind that certain work surfaces catch on fire when heated. Concrete floors make good work surfaces while working with a heat gun.**

oven. In seconds it produces excruciatingly hot temperatures, which makes it ideal for desoldering entire circuit boards all at once.

STUD FINDER

The last "power tool" that you will use is arguably not a power tool at all. The humble electronic stud finder is used to locate wooden support beams hidden behind your walls. Place it on the wall, turn it on, and slide it across the wall's surface until it beeps to indicate the presence of a wooden stud.

Stud Finder

Rulers and Tape Measures

Ever heard the old adage "Measure twice, cut once"? No advice could be truer. Before you make any precision cuts, you are going to need precision measurements. This brings us to the next set of tools that should be in your arsenal: rulers!

RULER

A two-foot metal ruler is a great tool to have, as it can be used as a straight edge for cutting and also used for measuring. A good ruler should have ⅛-inch markings on one side and ¹⁄₁₆-inch markings on the other. (If you prefer the metric system, look for a ruler that has millimeter and centimeter conversions on one of the sides.)

TAPE MEASURE

The other ruler you will need is a tape measure, a retractable metal or cloth ruler that can be pulled out to measure dozens of feet. This makes it ideal for measuring long distances, and because the ruler is also flexible, it can be used to measure around cylinders and other three-dimensional objects.

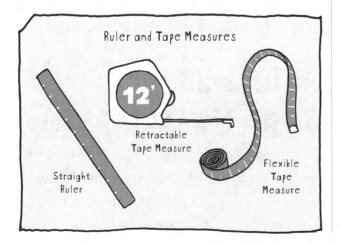

Ruler and Tape Measures

12'

Straight Ruler

Retractable Tape Measure

Flexible Tape Measure

Glue, Tape, and Fasteners

Moving slightly away from tools for a moment, there are other supplies that are extremely handy when it comes to building things. These include glues, tape, and fasteners. The one thing that all of these have in common is that they are typically used for bonding two things together.

GLUE

Adhesives come in all sizes and strengths. When it comes to glue, make sure you've got the right one for the job. Some glues work better on porous surfaces, some dry very quickly—know the properties of the surfaces you plan to glue together, and always, always follow the instructions for use on the packaging.

Craft Glue

The most standard glue that we will use is good old white nontoxic craft glue. As you know, this type of glue is not particularly strong and takes forever to dry. It works best when applied thinly and is good for gluing together paper and small craft notions.

Fabric Glue

Fabric glue is usually thicker and tackier than normal craft glue, which helps it join fabric (which will absorb the more watery white glues).

Hot Glue

Cranking it up a notch, we move on to the ubiquitous hot-glue gun. This is a true multipurpose tool and is capable of far more than just joining things together. The hot-glue gun can also be used to encase and protect something fragile, insulate electronics, and even make small leaks watertight. More often than not, hot glue is the only glue you will need, unless, of course, you need more strength.

⚡ **SAFETY FIRST**

>> As the name implies, hot-glue guns are hot! Keep a bowl of cold water around so if you get some hot glue on yourself, you can dip your hand in the water and limit the severity of the burn.

Epoxy

If you need something stronger, then you are looking to use a two-part epoxy. There are various different types of epoxies intended for fastening different types of materials, and you should always check to make sure the epoxy you are using will suit your needs. To use epoxy, mix it on a disposable surface (like a piece of cardboard) for the amount

⚡ **SAFETY FIRST**

>> Before using any glue or epoxy, read the safety warnings on the packaging.

Glues

Craft Glue   Fabric Glue   Hot-Glue Gun

Epoxy   Aquarium Glue

of time indicated on the packaging. Apply it to the surfaces you wish to join with your mixing stick.

As with many forms of glue, epoxy typically has a setting time of a few minutes, but it often takes many hours to reach full strength. Get in the habit of leaving something joined with epoxy to settle overnight.

Aquarium Glue

The one specialty type of glue that we will be using is aquarium glue. This is a silicon-based glue that provides a strong, gummy, watertight seal that can be used— you guessed it!—to make aquariums and anything else that needs to hold water. This glue is typically placed as thin beads between the joints of two flat surfaces.

TAPE

Sometimes you need to join things together, and you need to do it now! For this type of immediate application, you can use tape. Tape is essentially a strip of material (such as paper, fabric, plastic, or aluminum) with adhesive applied to one or both sides. Its tackiness and ease of use makes it highly versatile.

Scotch Tape

Scotch tape is the prototypical type of tape. It is a semitranslucent, one-sided tape, and we should have all been introduced to it at a very young age.

Two-Sided Tape

Similar to Scotch tape is two-sided tape. It has the same look and feel, except that two-sided tape has adhesive on both sides. This makes it ideal for sticking together two surfaces front to back.

Packing Tape

Moving up in scale, we come to packing tape. Since it is used for sealing boxes, it is stronger than Scotch tape and is typically crystal clear so people can see through it to read any important markings on the package. Its transparency makes it ideal for project applications when you need tape but don't want people to notice it.

Painter's Tape

Painter's tape is a paper-based tape that has adhesive on one side and tends to almost always be an opaque blue. It is essentially a special type of masking tape that does not leave behind residue on surfaces (nor does it tend to rip up dried paint when removed from walls).

Gaffer's Tape

Gaffer's tape is a strong, cloth-based tape that has a nice look to it and can be easily ripped lengthwise or widthwise. The other big advantage to gaffer's tape (as opposed to the ubiquitous duct tape) is that it almost never leaves nasty residue. Have a roll of gaffer's tape at your disposal if your hardware or craft store carries it.

Duct Tape

Duct tape is a strong, water-resistant tape that has more uses than I would care to list. It can do everything from fix ducts to make prom dresses and keep loose tennis shoes on your feet through class 5 white-water rapids. It's usually silver in color, but can be found in a wide variety of hues.

Aluminum Tape

Aluminum tape is basically aluminum foil with adhesive applied to one side. This type of tape is useful because aluminum is conductive and can be used to build crude circuits. Most hardware stores should carry this.

Electrical Tape

Conductive aluminum tape is handy, but usually you want tape that doesn't conduct electricity. Electrical tape is a stretchy PVC tape designed to wrap tightly around wires and keep them insulated. Electrical tape is typically black or red (but is available in other colors) and has a rubbery feel.

FASTENERS

Three common fasteners are used in this book and worth mentioning.

Screws

As alluded to earlier when we talked about twisting things with screwdrivers, we will, in fact, be encountering and using screws.

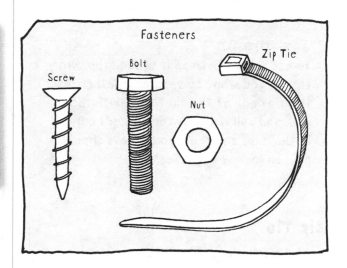

There are many different types of screws for fastening things to different surfaces. In this book, we will mostly be using wood screws and self-tapping screws.

Nuts and Bolts

Another very common type of fastener is the nut and bolt. Nuts and bolts come in a variety of sizes and are easy to come by. They are a strong and reliable way of fastening two flat surfaces together.

Tie One On
To use a zip tie, wrap it around the two things you want to secure together, insert the flat end of the tie through the boxy end, and pull it tight. You should hear a little "zip," and the two objects should now be locked together.

Zip Tie

The zip tie (or cable tie) is one of my personal favorite means of attaching things. These are quick, reasonably strong, reliable, and fun to use.

Techniques

Before you dive in, there are a few techniques you should know for making some of the projects in this book. Sewing and soldering are key skills when working with dead computers. Take a moment to get familiar with them.

SEWING

Working with fabric is an activity so special that it requires its own unique set of tools and skills.

The most basic tools you can use to work with fabric are a sewing needle, spool of thread, straight pins, a pair of scissors, pencil, and tape measure. With this basic arsenal of tools, you can accomplish an impressive

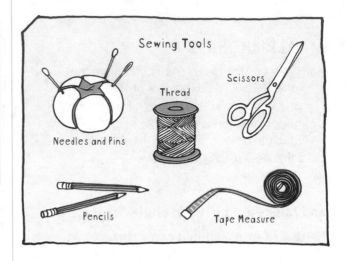

Sewing Tools

Needles and Pins

Thread

Scissors

Pencils

Tape Measure

amount of tasks. Of course, if you happen to like working with the heavy artillery—by this I mean a sewing machine, the long-range fixed gun emplacement of sewing—feel free. Though it will inevitably make things go faster, it is by no means necessary. Get a few hand-sewing stitches (page 34) in your repertoire and you're good to go.

Eyelet and Grommet Tools

Aside from the basics, you may also consider investing in an eyelet and/or grommet tool and the corresponding metal eyelets and/or grommets (sneakers often have these to pass shoelaces through). This tool is great for finishing small holes in fabric through which you need

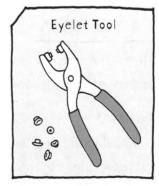

Eyelet Tool

to pass wires or bolts. Eyelets/grommets are also great because noncoated ones are conductive and can be used to attach wires to conductive fabric.

Seam Ripper

A seam ripper makes things all better. It's a small pencil-sized tool with a sharp hooked end for correcting stitching mistakes. This is particularly useful if you've been working

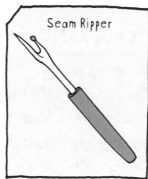
Seam Ripper

with a sewing machine—because of its speed, it's easy to make a "bad seam" very quickly. Luckily, you can tear out the messy stitches just as easily with a seam ripper.

Straight Pins

Straight pins are another helpful tool. They are used to temporarily hold seams together while you sew, keeping the fabric from getting misaligned. I prefer to use pins with large

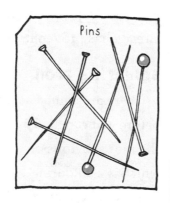
Pins

colorful beads at the head, because they're hard to overlook (and running a pin under the needle of a sewing machine can lead

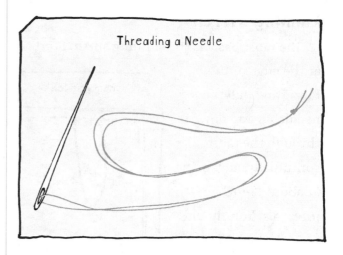
Threading a Needle

to disaster). Plus, if acidentally you spill a handful, they're easy to spot.

Thread the Needle

Start with the basics. Hand-sewing often gets a bum rap. People groan at the thought of having to hand-sew something because it brings to mind tedious and ineffective work. Total misconception. Hand-sewing can be not only enjoyable but highly reliable. Never underestimate the ability for hand-sewing to get the job done quickly and effectively.

To thread a sewing needle, insert thread through the eye and pull it through until you have a foot or two to work with. Cut the thread from the spool so that you have an equal amount to the length you just pulled through. Fold the thread evenly in half and tie the ends of the thread together in an overhand knot. Once the needle is threaded, move on to some basic stitches.

Running Stitch

The most basic is the running stitch. Start at the edge of the fabric and pull the needle up and down through the fabric continually, advancing at about ⅛-inch intervals. You should see a dashed line start to form. Continue to do this until you have completed the seam.

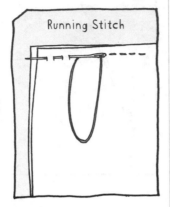

Whipstitch

The next stitch and one employed most frequently throughout the book is my personal favorite, the whipstitch. I like the whipstitch because it is quick and effective. If you don't believe me, take a needle and pull it through both layers of fabric from the back on up. Advance along the edge of your fabric slightly, drop the needle and thread back below the fabric edge, and pull it through again. Continue doing this until the seam is sewn shut. (That should all have been very easy and effective, proving my point.)

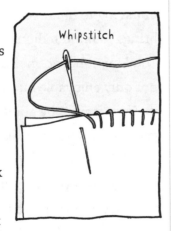

Slipstitch and Buttonhole Stitches

Other stitches you will use are the slipstitch and the buttonhole stitch. These will be covered in Resistor Pillow and the Keyboard Buttons projects, respectively (pages 166 and 104).

SOLDERING

Soldering is a large component of computer manufacturing. It should come as no surprise, then, that soldering is also employed in repurposing old computers. Often the best way to transform a dead computer into a new form or state is to disassemble and inventively re-create it using soldering techniques. The nice thing about soldering is that it is structural, functional, and, as the circuit board already illustrates, practical. We know it will work in joining electric components together because it already has proven itself.

Soldering Iron

There are some basic tools you need before you can solder, the most obvious of which is a soldering iron. As a beginner, you may want to purchase a cheap fixed-temperature

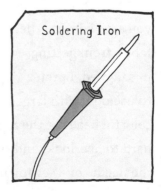

soldering iron from your local electronics store (this can be 15 to 40 watts). As you get more advanced, I recommend you buy an adjustable-temperature soldering iron so you can vary the temperature for different soldering jobs.

Solder

The next extremely obvious thing you are going to need is solder. Solder is a metal alloy with a low melting point, and it's used to fuse two metal surfaces together. There are two things to keep in mind about your choice of solder: the gauge of the

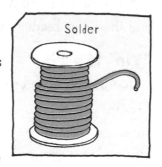

wire and the material. If you plan on soldering large things together, you should use a larger-gauge solder. However, if you are soldering small things to circuit boards, a smaller-gauge solder is better.

For our purposes, there are two types of solder: those with lead and those without lead. I recommend that you try to use lead-free solder. However, lead-free solder often produces more nasty fumes than lead solder, as it has more flux in it and usually starts to melt at higher temperatures. If you have a respiratory condition like asthma, you may consider using lead solder, as the fumes, albeit

still harmful, are slightly less dangerous (you just have to be cautious about exposing your skin to lead).

Other Soldering Materials

Before you can start, you need a soldering iron holder so you can put your iron down and have it held safely in place so you don't burn the tabletop or yourself.

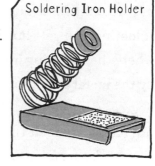

You will also need a damp sponge for cleaning off the solder tip between soldering.

Other things you may consider having on hand are extra solder tips in case the one in your soldering iron gets too dirty to use, and various metal picks and knives for poking and scraping at solder joints. As a beginner, you should have some desoldering braid (a copper braid for

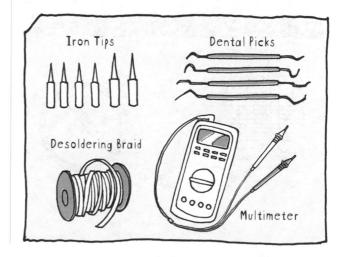

removing solder from a soldered joint) for cleaning up your mistakes. Lastly, to test whether or not you have made mistakes, you need a multimeter with a continuity tester mode.

Soldering Safety

Before you start, it is important to take into account some safety issues. Foremost, the soldering iron gets hot. Always be mindful of where the soldering iron is, and never grab it by the metal part, as it can get as hot as 800 degrees (Fahrenheit). This temperature is hot enough to melt through plastic, and you should also be mindful of where the soldering iron power cord is so you don't melt through it and electrocute yourself.

The other safety concerns are a little less extreme. You should always use an exhaust fan to blow the smoke away from you and

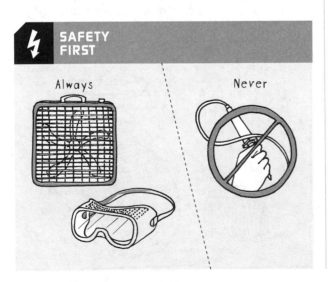

SAFETY FIRST

Always   Never

out a window (or some comparable means of ventilation). A standard box fan usually should suffice. Use safety goggles, as molten solder can sometimes "pop" and get into your eyes. In addition, solder fumes are an irritant to your eyes (which can be reduced by using an exhaust fan).

Once you have set up your soldering station, turned on your exhaust fan, and lightly dampened your sponge with tap water, you should be ready to start.

"Tin" the Tip

Before you do anything, it is a good idea to place a little bit of solder onto your new soldering iron tip. This is called "tinning" the tip. Turn on your soldering iron and wait for it to reach a high temperature. Pick up your soldering iron by its insulated handle as though you

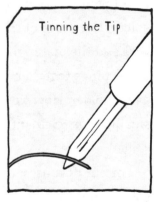

Tinning the Tip

are holding a spoon. Using your other hand to hold the solder (so the end of the solder wire is a few inches from your hand), carefully melt a small amount of solder onto the surface of the tip. After you accomplish this, then wipe off the tip on the damp sponge and place it in the holder. It may hiss, smolder, and produce a little steam, but don't worry, this is normal. Now you should be fully ready to start soldering!

connecting Wires

The easiest form of soldering involves connecting two wires together. Wires are easy to connect because you can heat them a great deal without having to worry about them overheating (as they are just thin metal strands).

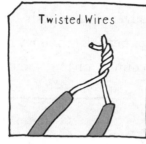

Twisted Wires

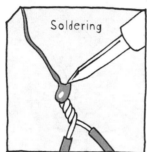

Soldering

To solder wires together, first strip back a little bit of insulation from the ends and either twist or hook them together.

You then firmly pick up your soldering iron, again, like you are holding a spoon. Carefully place the tip of the iron on the joint of the two wires and heat it up for a few moments. After a second or two, use your other hand to push some solder into the heated joint of the wires (close to the tip of the soldering iron).

Remove the soldering iron and the remaining solder from the wire joint and

watch as the solder quickly changes from appearing fluid to solid. Wipe the tip of the soldering iron on the damp sponge to remove any dirt or extra solder.

Cleaning

HELPFUL HINT
Cleaning the soldering iron tip on the sponge after every use is important to keep the tip clean and working properly.

connecting to a circuit Board

Soldering becomes slightly trickier when connecting things to a circuit board. Although there are only a few projects in this book that connect parts to circuit boards, this is good information to know as it is a fundamental skill of working with electronics. Like many basic skills, it isn't too hard to get started, but it can take a lifetime to truly master it.

HELPFUL HINT
You may be tempted to melt the solder onto the tip of the iron and then place it on the wires. You should avoid doing this, as it can result in bad solder connections.

For this book, you need only be concerned

with "through-hole" soldering. This means that you will be inserting electronics through holes drilled in the circuit board and soldering them to solder pads on the backside.

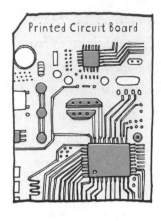

Printed Circuit Board

To do this, first insert your electronic component and make sure that it is going through the proper hole to the proper metal solder pad. Once you are sure that everything is correct, firmly grasp your soldering iron and heat up the joint between the component's wire lead and the metal solder pad. You don't want to heat this up for too long, as excessive heat can damage electronic components. The goal is to quickly place the solder iron tip at the joint while simultaneously inserting solder into the joint

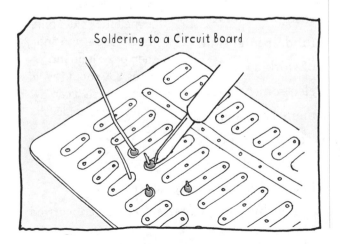

Soldering to a Circuit Board

until it melts just enough to coat the metal pad and cover the hole. Again, when you are done, clean the solder tip on the sponge and put the iron in the holder.

The resulting solder joint should look shiny, silver, and smooth. If it looks jagged, dull, or beadlike you may have made a bad solder connection and should fix it.

Comparison of Solder Joints

If the connection is bulbous, try lightly picking the solder off with one of your metal scraping tools (like a dental pick). There is a chance it may fall right off. Otherwise, you will need to use your desoldering braid. Place a section of desoldering braid atop the bad solder connection and then heat it up using the tip of your soldering iron until the solder starts to melt into the braid. Remove both and clean the tip of the iron. Wait for the joint to cool off and try soldering the joint again with new solder.

Solder joints need to be kept separate. If two joints become connected by solder, there is a great chance that your circuit won't work, because electricity always tries to find the path of least resistance—by

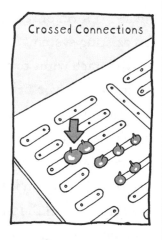

Crossed Connections

making extra connections, you are changing the path in the circuit and therefore both the direction in which electricity will flow and how the circuit will operate.

HELPFUL HINT

The only exception to the above scenario is if the two joints share the same solder pad. If they do, then they will obviously be connected by solder.

If you're concerned that two solder pads on the circuit may have accidentally been connected, test them with your multimeter to make certain (please see page 19 for an explanation on how to do this). Once you've confirmed that two

Multimeter Testing

solder pads are connected (and shouldn't be), proceed to the desoldering section, below.

Desoldering

If you've determined your circuit has a short in it, you can try to break the solder connection between the two pads manually using a scraper tool. This is usually ineffective, however, and may damage the circuit board if too much force is applied, but it is always worth a quick, and gentle, try.

If manually breaking the connection doesn't work, place desolder braid on top of the solder that is bridging the two pads and heat it up with the tip of a soldering iron until the solder starts to melt. Once the braid has absorbed some solder, remove it, clean the tip of the soldering iron on a sponge, wait for the pads to cool, and then test the continuity again. If the meter does not beep or change from 1 to 0, then you have fixed it. If it still does, repeat this process until the pads are no longer connected.

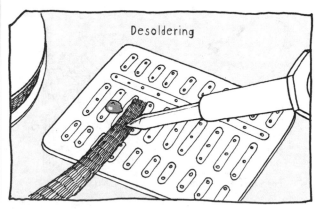

Desoldering

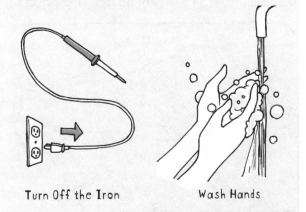

Soldering to a Circuit Board

Hacking a Circuit Board

There are a few times in this book where we will need to attach wires to terminals on existing circuit boards that have been taken out of dead computers (as opposed to building them ourselves). This is called a "hack," a form of reworking or circumventing an existing system.

To attach wires to solder terminals on a circuit board, you first need to "tin" the wire you would like to attach by coating it with a small amount of solder. Once coated, place this wire atop the solder pad on the board you are trying to attach it to. Hold the soldering iron tip on both the tinned wire and solder pad until they melt together. Quickly remove the tip and wait for the solder to solidify and hold the wire in place. If the wire falls off, wait for it to cool down and try again. Clean the tip of your soldering iron when you are done by wiping it on the sponge.

⚡ SAFETY FIRST

>> When you are done soldering, don't forget to turn off the power to your iron (or unplug it). Before you do anything else, it is very important to wash your hands well with soap and water (especially if you were handling lead solder).

Turn Off the Iron   Wash Hands

Safety

Dead computers and just about every other electronic device are usually classified as toxic waste. As such, you are bound to encounter a number of potentially horrible things while repurposing them. I can go on all day listing things that may potentially be harmful to the environment and, worse, yourself. Rather, I'll just go over some basic concerns and general rules of thumb.

The older the electronic device, the more likely it is to have fewer consumer protections and contain highly toxic and dangerous materials. In addition, the larger the device, the more potential it has to contain things that may badly shock (as in, electrocute) you. It would follow that an old TV is potentially more harmful than a newer MP3 player.

As a general practice, avoid heating, grinding, sawing, bashing, and generally disrespecting the item you are working with. Avoid disassembling basic components such as transistors and microchips.

Silicon, Mercury, Beryllium Oxide, and Cadmium

I cannot emphasize enough that grinding, sawing, and dismembering things that you cannot identify (and some things you can) is a horrible idea. For instance, you may be tempted to cut up circuit boards into all kinds of shapes and sizes. However, doing so improperly may lead to a degenerative, incurable lung disease called silicosis (similar to asbestosis) later in life. You may also encounter a number of really toxic substances (mercury, beryllium oxide, cadmium) hidden within computer components, but only if you decide to cut, grind, or saw them open. Exposure to these nasty materials can be avoided by not sawing, grinding, cutting, or otherwise modifying basic electronic components.

Almost every component you will encounter is benign if the casing is left undisturbed.

Lead

The one heavy metal you will come into repeated contact with is lead. All circuits made prior to 2003 (and a good deal made after) almost assuredly contain some amount of lead, a heavy hitting heavy metal—it's known to cause birth defects, other reproductive harm, and cancer. Lead is used in the production of a number of computer parts, and is also found in most solder. It's good practice to always wash your hands with soap and water after handling circuit boards and/or solder.

Electricity

Aside from hazards posed by the materials themselves, danger also lurks where we can't see it. Many old (seemingly dead) circuit boards still store dangerous amounts of electricity. Any amount of electricity with a high voltage (over 15 volts) or high current (1 A, or 1000 mA, or greater) is potentially harmful. Note: When dealing with capacitors, the dangerous ones to avoid are those rated higher than 25 volts

SAFETY FIRST

>> Always make certain that electronic devices are unplugged before you start disassembling them.

and 1000 μF (these ratings are printed on electrolytic capacitors).

It is safe to assume that any device that plugs into a wall stores an electric charge that is large enough to harm and potentially kill you. The source of this charge is a large high-voltage capacitor usually located near the power plug. You can identify these capacitors because they usually have a voltage rating written on them of 100 or more volts.

Whenever you have a circuit board containing a capacitor rated at 100 or more volts you should discharge it. If you can get away with not discharging it, I recommend that as option number 1. However, if discharging it is absolutely necessary or it would make you feel better to know that it is discharged, see page 13 for a method to do so. Keep in mind that it is more of a "hack" method than the professional way to do things, but it is effective if performed safely and correctly.

Computer and Television Monitors

Computer monitors are possibly the most dangerous device you may encounter as a casual hobbyist. They can possibly shock, radiate, and poison you. And on top of that, they have been known to violently implode, sending glass and toxic chemicals flying around with thousands of pounds of force. I tell you this primarily to scare you. The moment that you stop being scared of monitors is when potentially lethal accidents happen. If you have any doubts about working with such dangerous material, don't do it. There are many organizations that will happily recycle an old CRT (cathode ray tube) monitor—look up your options at the local recycling center.

Here are some suggested guidelines to follow when you're working with a CRT monitor:

1. Always work with someone else present who can call for help in the case of an emergency. A large electric jolt is quick and unforgiving.

2. Always wear thick rubber gloves, approved shatterproof safety goggles, and shoes with rubber soles.

3. If possible, work with your left hand in a pocket or held behind your back. You may think this sounds silly, but this could save your life. Electricity always travels to ground through the path of least resistance. If you use both hands, the electricity may travel through one arm and out the other,

> **SAFETY FIRST**
>
> >> **Cut off the wall plug of anything you are working on to prevent any future mishaps.**

in the process passing through the heart and causing sudden cardiac arrest. If you only use your right hand, there is little possibility that electricity will pass through your heart and cause cardiac arrest. That said, the chance that you get shocked is still *very high*.

4 Respect the tube inside of your monitor. The cathode ray tube is a vacuum containing roughly 14 pounds of force for every square inch. That means the tube has the potential to implode with many tons of force. Do not hit or drop it. Be particularly careful around the skinny part of the neck, since this is the part most likely to break. Never, and I repeat, NEVER, pick it up by the skinny part of the neck. Always carefully lift it from the front viewing screen where it has the most surface area. Again, if the monitor implodes and you happen to be too close,

it can kill you. Should you survive the implosion, the doctors will have a field day picking all of the shards of glass coated in toxic chemicals out of your wounds. Yes. This is scary stuff. If you have any doubts about working with such dangerous material, don't do it.

5 Once you start to disassemble a television, never, EVER, plug it back in. In fact, cut off all wall plugs before you unscrew a single screw.

Remember

Safety issues are not to be taken lightly, and it's important that you have a thorough understanding of them in order to approach your next DIY adventure with reverence and respect. Please also read the Notice to Readers: A Brief Word on Safety on page ii before you dig in. Be safe, use common sense, and have fun.

SAFETY FIRST

>>Some televisions monitors have capacitors that store many thousands of volts. *Never, ever* discharge one of these capacitors with the Christmas light technique as shown on page 13. I highly recommend avoiding handling these capacitors altogether, if possible. However, if you must handle it and need it discharged, I have found that throwing a ball of aluminum foil at the solder terminals from a safe distance (at least 1 foot) to be highly effective. You will know it is being discharged when you see (and hear) a large spark. You should then throw it a dozen more times for safety's sake—to make sure. If you do this many, many, times and nothing happens, put on a rubber glove and, using your right hand, try bridging the terminals with a long insulated screwdriver. If still nothing happens, it is probably safely discharged—but always use caution nonetheless.

CHAPTER 2

Projects for a Postconsumer Dwelling

> **Get smart with 14 projects to deck out your desk and office, live large in your living room, and geek out your kitchen with your fried electronics.**

There are two types of waste in this world, preconsumer waste and postconsumer waste. Which one are you? No, wait. Don't answer that! You are not waste. You are a unique, talented, and special individual. Your broken printer, on the other hand, *is* waste. The difference between preconsumer waste and postconsumer waste is pretty self-explanatory.

Preconsumer waste is the material left over after the manufacturing of consumer products; it is discarded before it has any consumer application. This type of waste, albeit serious, is not as problematic as postconsumer waste. The reason is that the goal

Mr. Resistor Man Says:
As of June 2008, there have been more than 1 billion personal computers distributed worldwide.

PROJECTS

of any business, generally speaking, is to keep the cost of products as low as possible—since raw materials cost money, it's ideal to minimize this type of waste from the onset. On top of that, production scraps are often reused in the creation of more products (as another type of money-saving measure). This thriftiness on the part of the manufacturers is good for business and the Earth.

Postconsumer waste is the waste generated when a product reaches the end of its life cycle and can no longer fulfill its intended use. Whereas the factory has an incentive to recycle or reduce waste in production, the consumer's incentive (that's yours) is pretty low. First, reducing consumption of products in the first place is hard (and not exactly encouraged by profit-seeking companies). Second, in most places, it's a pain to figure out how best to recycle waste—especially electronic waste (which often requires the consumer to seek out a specialty service performed at a fee).

Sustainable Vase, p. 77

Scanner Compost Bin, p. 63

cable cord coaster, p. 50

Magnetic Memo Board, p. 70

The sad truth is, there's little stopping us from throwing away our broken consumer electronics. Two solutions: 1) Buy less junk (not likely, judging by how we clamor for the next "it" gadget) or 2) repurpose our broken electronics. A broken scanner that is converted into a table is no longer a scanner, but a piece of furniture. It has been transformed from its initial state and taken on a new purpose—making it no longer postconsumer waste.

In high school I was taught that cultural change starts in the home. Let's take this message to heart and, without fear of being too literal, start with projects that repurpose broken electronics for use in our homes. Let us shape our homes to be a little more Earth-friendly and tech-beautiful, and watch and wait as the world follows.

#1 SCANNER Table

LEVEL 2: Intermediate | **TECH TRASH:** Scanner

I sure do like things. Things like windup triceratops toys, paperback novels, medicated foot creams, shiny pipe cleaners, and badly framed portraits of strangers. Now, if there were but one thing that I like more than anything else, it would be a small end table to place all of these fabulous things on. If that table were sleek, stylish, and easy to make out of recycled material, then it would be perfect. Well, what do you know? This scanner table is all of that and more—you even fill the large empty space of the scanner bed however you like!

MATERIALS

- Scanner
- Phillips-head screwdriver
- Drill with a 3/8" drill bit
- Four 3/8" by 18" steel rods
- A hex wrench set
- Eight rod clamps
- Four 1/8" cable fasteners
- Ruler
- Steel cable
- A concrete surface
- 1" chisel
- Hammer

Mr. Resistor Man Says:
The first image ever scanned was a personal photograph of the three-month-old infant son of the man who led the team to the scanner's invention.

MAKE IT

1. GUT THE SCANNER

Flip the scanner over and remove the screws fastening the top and bottoms together. Then remove all of the electronics and mechanical assemblies from the inside of the scanner so that you are left with an empty shell.

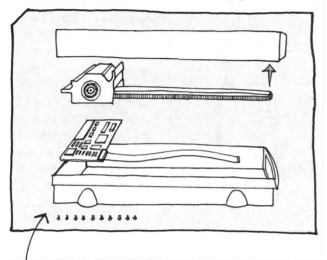

Set aside these parts for later reuse in projects such as the Gear Clock (page 56) and Mr. Resistor Man (page 129)!

2. DRILL HOLES

There should be four rubber pegs on the bottom of the scanner. Remove them by prying them out with the screwdriver. Then drill a ⅜" hole in each of the four bottom corners where the pegs used to be.

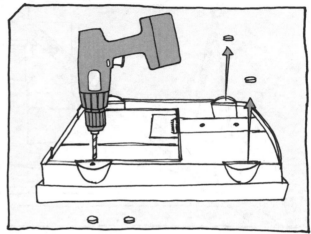

3. ROD CLAMPS

Attach the rod clamps to the rods so that they are all attached 3" from the end of the rod. Actual measurements may vary depending on the depth of your scanner (you will make adjustments later by trial and error).

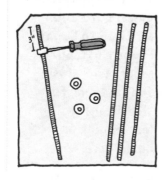

4. INSERT THE TABLE POSTS

With the scanner still "face" down, insert one of the legs, rod clamp end first, through one of the drilled holes. Push it until the rod clamp rests against the base of the scanner. Repeat with the three remaining table legs. Adjust the location of the four rod clamps as needed.

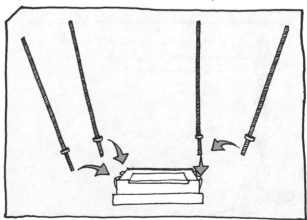

>>>>

5. MEASURE

Carefully (so the rods don't fall out) right the table. Measure diagonally between the center of each set of rods on the top inside of the table. This measurement (x) will be used to create tension cables.

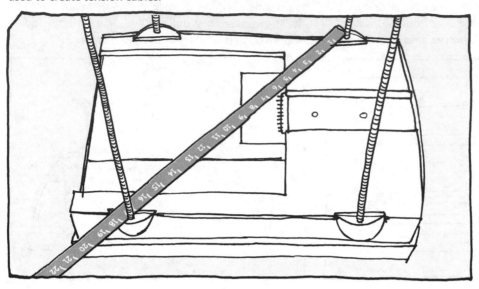

6. PREPARE THE CABLES

Pull a few inches of steel cable through a cable fastener, fold it, and insert it back through the fastener to form a small loop. After making the first loop, insert another fastener onto the wire and create another loop. From one end of the first loop to the end of the second loop, it should measure between 1/16" and 1/8" less than the diagonal measurement (x) you made in the previous step.

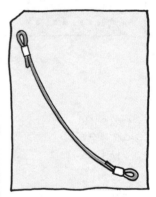

7. HIT IT

Place your fastener (with looped cable) on a concrete surface. Center the chisel edge perpendicularly over the fastener (so that it crosses both cables inside). Firmly grasp the chisel and give it a few good whacks with a hammer, crimping the cable inside the fastener. Repeat on either side of the original notch.

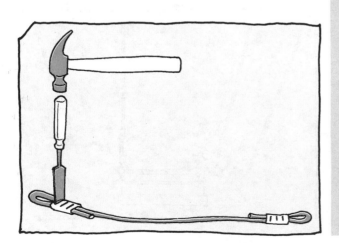

Save it!
The tool used to properly clamp shut a cable fastener costs upward of $200. Feel free to buy it if you must, but I recommend closing the fastener for significantly less money, as noted here!

SAFETY FIRST

>> Be careful to keep your fingers out of the way of the chisel and hammer. Your fingers are important!

8. ATTACH WIRES

Slide one of the cable loops around one of the table legs. Slide the loop at the opposite end of the cable around the opposite leg on the diagonal. Attach the second cable over the remaining legs so it forms an "X" on the inside of the scanner. Hold them in place by fastening rod clamps onto each post directly atop the wires.

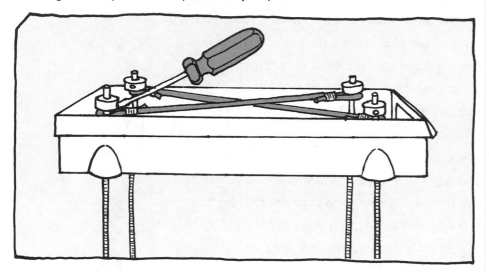

9. PERSONALIZE

Here is your opportunity to let some personality shine through. Decorate your scanner by placing objects of your choosing (yarn, tools, etc.) inside the casing.

10. REASSEMBLE

Reassemble, popping or screwing parts of the scanner back together so the scanner bed is firmly attached to the base.

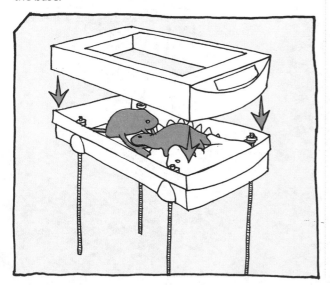

HELPFUL HINT

Note: This table is not intended for heavy load bearing. Don't sit or stand on it or use it to hold heavy things like cinder blocks or elephants. (Remember, it's still fundamentally a scanner!)

#2 CABLE CORD Coaster

LEVEL 1: Novice | **TECH TRASH:** Computer Power Cable

A while ago I had a box full of random cables in my life. Now, I have three boxes. Perhaps, as a junk collector, I may not be the ideal case study for illustrating my point. However, even if my proclivity for amassing random wires may make me an oddity, I will venture to guess that you have a few stray cables cluttering up your life, too. There is hope yet for that collection of tangled confusion you have amassed.

When life hands you an existential tangle of wires, make orderly spiral coasters out of them. It says that you are a person capable of taking the chaos surrounding you and turning it into harmonious order. Indeed, coasters are the true mark of self-realization. You can quote me on this.

MATERIALS

▸ Large plastic cup
▸ 1' sheet of cork (or place mat size)
▸ Marker
▸ Scissors
▸ Diagonal cutters
▸ Computer power cable
▸ Gaffer's or duct tape
▸ Hot-glue gun

MAKE IT

1. TRACE A CIRCLE

Place the cup upside down on the cork mat. Trace a circle around the lip of the cup and cut it out with a pair of scissors.

2. SNIP

Cut off the end of the power cable that normally would connect to the computer. Cut an additional 2" piece of cable from the end and set it aside for later.

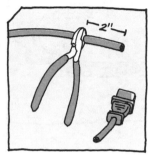

3. MAKE A TAPE SQUARE

Cut two approximately 4"-long pieces of tape. Lay them sticky side up next to each other to form a square on your work surface.

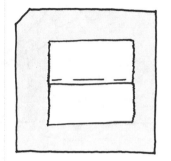

4. THE INITIAL LOOP

Coil the trimmed end of the cable into the tightest possible loop that you can make. Stick it firmly onto the center of the tape square.

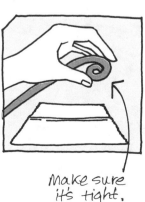

Make sure it's tight.

5. COIL IT

Continue to tightly coil the cable around the center loop until you have created a spiral just slightly bigger than the cork circle. Your coil should be stuck firmly to the piece of tape. Use the 2" piece of cable you cut in Step 2 to plug the opening in the center of the coil.

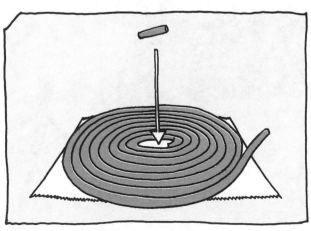

6. GLUE

Liberally apply hot glue to the surface of the wire coil. While the glue is still hot, center and press the cork circle over the coil. Hold it firmly in place until it dries.

7. TRIM

Cut off the excess cable where it starts to spiral out from under the piece of cork. Glue the end of the cord and hold it in place until dry.

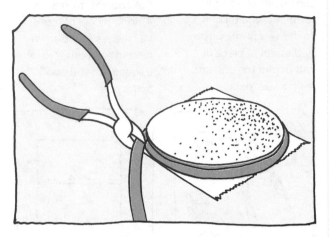

8. PEEL

Flip the coaster over (so the cork side is down) and gently peel off the tape. If needed, fill the center of the coaster with hot glue to further seal it.

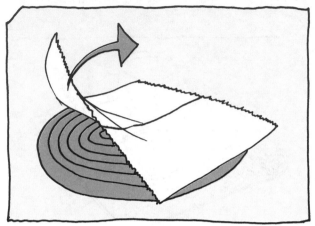

A set of coasters makes a great gift for the home.

Mr. Resistor Man Says:
To remove cup rings from wooden surfaces, place a dishtowel over the ring and with a dry iron, move back and forth over the towel for 20 to 30 seconds. Then make these coasters so it will never happen again!

#3 Wire LANTERN

LEVEL 2: Intermediate | **TECH TRASH:** Large Wires and Cables

Generally speaking, I like my home decor simple. So when I had the sudden urge to haphazardly tie together a jumble of wires to form this spherical lamp, it was highly out of character. But when I popped the balloon at the end, I was pleasantly surprised with a delightful lamp that surpassed all of my initial expectations.

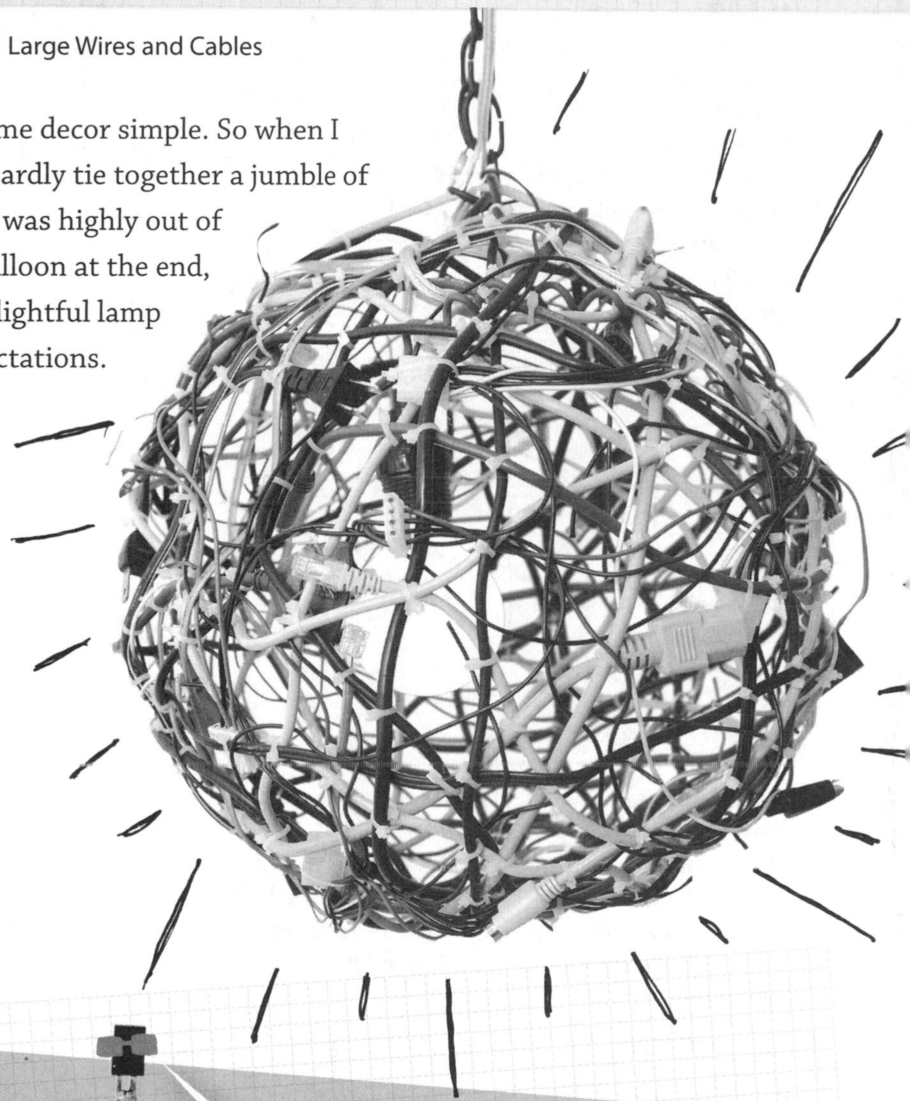

MATERIALS

- ▶ Stud finder
- ▶ Power drill
- ▶ Two 3" ceiling hooks
- ▶ Scissors
- ▶ 5 yards string
- ▶ 36" balloon
- ▶ Box full of large wires and cables
- ▶ Pile of small wires and connectors
- ▶ 400 4" zip ties
- ▶ Wire clippers
- ▶ Hanging lamp socket (preferably 12 or more feet of cable)
- ▶ Lightbulb
- ▶ 3' to 5' strong rope or chain (long enough to hang comfortably from the ceiling)
- ▶ 1/8" quick link

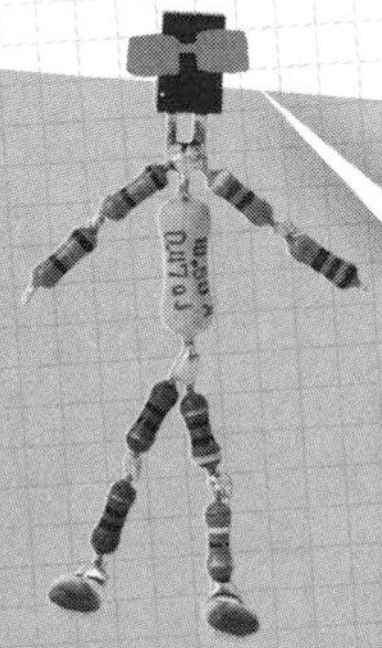

Mr. Resistor Man Says:
By 2007, roughly 235 million electronic devices had been tossed into storage in America alone.

MAKE IT

1. THAT'S THE SPOT

Use a stud finder to find a ceiling beam. (*Note:* If you don't have a stud finder, you can do this the old-fashioned way by tapping on the ceiling until it "sounds right.") Mark it and drill a pilot hole into the ceiling. Then screw a ceiling hook into the beam.

Note: To avoid putting any extra holes in your ceiling, make the lantern in the same place that you want to hang the final thing.

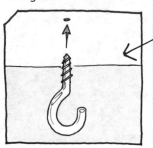

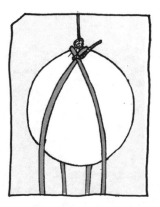

2. LOOPY

Cut a piece of string equal to the distance from floor to ceiling. Tie a loop at one end and attach it around the ceiling hook.

3. KNOT IT!

Blow up your balloon, knot the end, and tie it with the other end of the string at about shoulder level. Trim off any excess string.

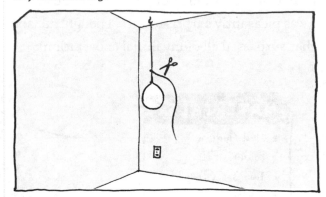

4. DRAPE

Drape two large wires over the balloon, crisscrossing them at the top. Use a zip tie to attach them together at their intersection. Continue draping and tying large wires in this manner. Once a handful are attached at the top, start to wrap their ends around the bottom and sides of the balloon in a web and connect them at every intersection.

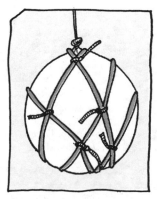

5. SNIP AND WEAVE

Using your clippers, cut off the extra plastic from the ends of the zip ties. Weave your smaller wires in between the larger wires already attached around the balloon. Zip-tie the smaller wires only where necessary to secure them.

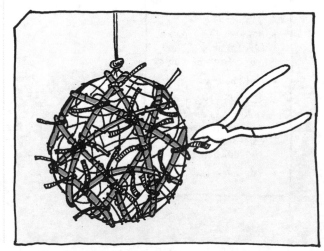

6. POP THE BALLOON

Grasp the wire cage so that it won't fall, and pinch the balloon at the knot. Make a small incision to deflate the balloon.

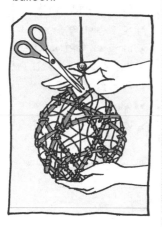

7. ADD THE LIGHT

Insert the socket end of the lamp cord through the opening at the top of the wire cage. Weave the wire across the gap on the top of the lamp, and then drape the socket over this wire into the center. Insert zip-ties to secure the intersections of the wires. Next, weave the cord in and out once around the circumference of the lamp, all the while securing it in place. Screw in the lightbulb.

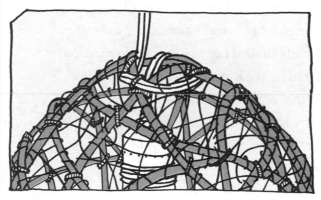

8. CHAIN

To attach the chain, pass it across the opening in the top of the lamp, through two strong sections of wire. Lock the chain to itself using the quick link, as shown.

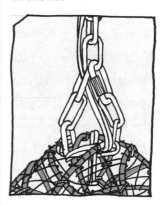

9. HANG

Hang the lamp from its chain around the ceiling hook. Wrap the power cord around the chain and drape it, too, over the hook. Decide where the lamp will plug in and add an extra ceiling hook above the socket; stretch the cord over the hook, down the wall, and plug it in. Let there be light!

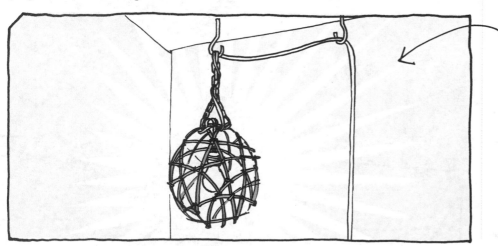

A clear, 25-watt bulb is a nice alternative to install in this lantern.

#4 Gear clock

LEVEL 1: Novice | **TECH TRASH:** Printer

If I were a clock, I think that one of the most terrifying things that could happen would be seeing my own gears spilled out all over the floor in front of me. I imagine that this is the clock equivalent of having to hold a pile of your own guts in your hands. Chilling.

If, however, it's someone else's guts we're pondering over, the horror tends to gradually shift toward fascination. We are curious creatures and easily captured by the inner workings of complex systems—we study skeletons, charts, and renderings of the human body in an effort to understand. I'd venture that a clock feels the same way about its own gears. In order for it to fully understand its own nature, it has to come to terms with the functionality of its inner mechanisms. And once put into this perspective, I see little reason why a clock would not want to adorn itself in exaggerated gears.

MATERIALS
▸ Printer
▸ Cheap wall clock
▸ Screwdriver
▸ Hammer (optional)
▸ Hacksaw (optional)
▸ Pliers
▸ Diagonal cutters
▸ Wrench
  ▸ Hot-glue gun

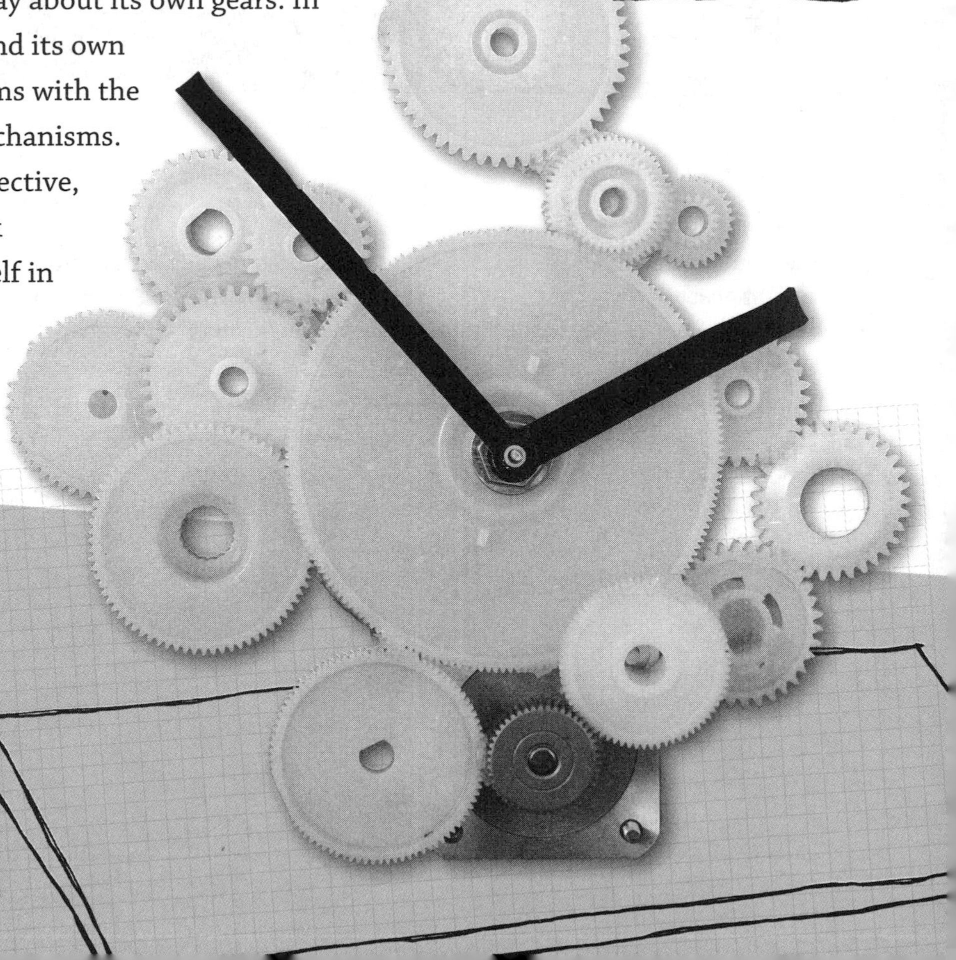

MAKE IT

1. REMOVE THE COVER

Remove any screws and pry open the printer's plastic covering. (You might need to use some force.)

Don't be shy about using that hammer and hacksaw to get that printer open.

2. GEARS

Find the printer's gear-drive mechanism. (This is usually on the side of the printer opposite the ink cartridges.) Remove the gears from the printer by gently prying them off their shafts with a screwdriver (first you may need to remove clamps that are locking them in place).

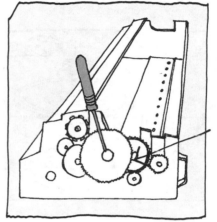

This gear is ideally large enough to completely hide your clock movement.

3. MOTOR

There will be a large motor with a gear attached to it that was driving the gear assembly. Remove it from the printer to serve as the base and set it aside.

HELPFUL HINT

The second hand is often glued onto the clock's drive shaft. Disassembly may require cutting the second hand free which will mean that you will no longer be able to reattach the second hand (which is okay—life is so much more relaxing when you don't watch the second hand anyway).

4. CLOCK

To remove the clock mechanism from the wall clock, first remove the clock hands and unscrew the nut that fastens the mechanism to the clock face. Set aside the clock hands and any additional hardware for reassembly.

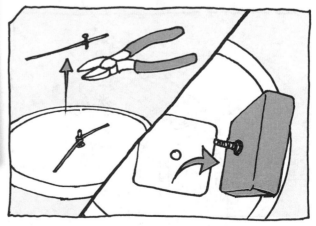

>>>>

5. SIZE IS EVERYTHING

Find the largest printer gear. With your diagonal cutters, snip away any plastic that sticks out and prevents your clock mechanism from laying flush against the gear.

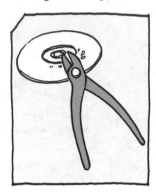

6. ATTACH THE CLOCK

Attach the clock movement by passing its main shaft through the center of the gear. Securely fasten it using the hardware you removed in Step 4.

7. STEADY IT

The motor will now serve as the base for the clock. Apply a fair amount of glue between the gear and the motor so the gear can no longer turn.

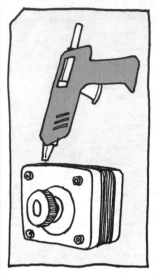

Note:
If your clock motor is round rather than rectangular, glue the motor to a flat base with a large surface area, such as a thin piece of metal or plastic.

8. SPACING

Glue a small gear to the gear on the motor, as shown. Reinforce it by gluing the gear to the motor, too, to hide the point of contact. This first gear should serve as a base to attach the larger gear that hides the clock mechanism in Step 9. As such, this gear should be positioned so that it rises above the top of your motor.

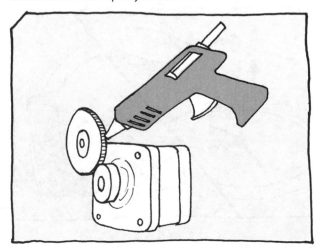

9. LARGE GEAR

Glue the large gear with the clock movement to the spacing gear you attached in Step 8 so that the clock shaft is aligned directly over the center of the motor.

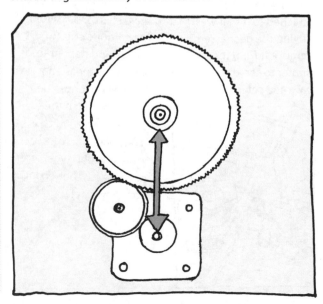

10. STEADY THERE

Glue on another gear in such a way that it makes contact with the clock movement, the large gear, *and* the body of the motor. Be sure it is secure, since it will provide stabilization for the entire clock. A gear with hub or shaft is ideal, because the part that will be making most of the contact will end up hidden from view. Continue hot-gluing gears, without overdoing it, until you like the way your clock looks.

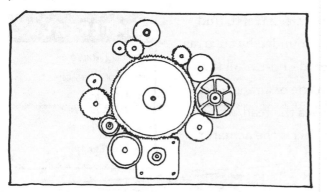

11. GIVE ME A HAND

Reattach your hour and minute hands to the clock's drive shaft.

12. POWER

Insert a battery into the movement to power it up. Set the clock to the proper time.

(back)

GEAR CLOCK

>> Variation

Before you glue all the gears together (or, you can use the gears left over) for the clock, make a gear-pressed notebook. Trim the hubs off the gears so they are completely flat and arrange them, "clean" side down, against the cover of a moleskin notebook. Sandwich your notebook and gears between two pieces of wood and clamp your sandwiched notebook using a table vise rotated parallel to the floor plane. Attach a C-clamp or two if you have them to provide uniform pressure over the surface of the wood. Wait 30 minutes. Remove the clamps from your notebook, shake off the gears, and—voilà!—an elegantly textured notebook to inspire.

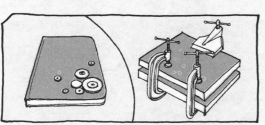

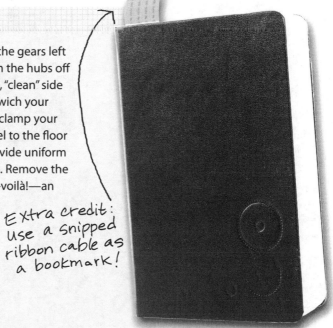

Extra credit: use a snipped ribbon cable as a bookmark!

#5 Actuator Arm MAGNETIC Picture Display

LEVEL 1: Novice | **TECH TRASH:** Hard Drive

I t may sound futuristic, but you don't need to time-travel to find an impressive actuator arm. In fact, you're surrounded by an array of actuator arms all the time. (So put that cardboard-and-tinfoil time machine back in the closet and let me explain.) Inside of almost every computer hard drive, there is a little arm that moves the read/write heads back and forth over the hard drive platter. This is the actuator arm. If you consider the amount of devices with hard drives in them, actuator arms are everywhere. The future is now!

Mr. Resistor Man Says:
Hard drive capacity has been increasing at 40 percent a year.

MAKE IT

1. VOID THE WARRANTY

Use your screwdriver and/ or Torx wrench to carefully remove the cover from the hard drive. (This, FYI, will void the warranty. Your hard drive is officially dead now.)

2. PRY OFF THE MAGNET

Find the magnet covering the back part of the actuator arm. Remove any screws holding it in place, and then use your flathead screwdriver to carefully pry the top magnet away from the bottom magnet. These magnets are very strong and this action may take a lot of force. Just be careful not to break the magnet or free it too forcefully (you may send it flying across the room!).

Heavy Duty

Set aside your wallet while working with these large magnets.

They are very powerful, and if you accidentally drop one in your lap, you might inadvertently erase all of your credit cards!

3. REMOVE THE "PLATTERS"

There should be a series of screws clamping the platters to the spindle. Remove the screws and the plate that is clamping them down. Rotate the actuator arm so that it is no longer sandwiched between the platters. Finally, remove the nice shiny platters and set them aside for future projects (like the Obligatory Decorative Candleholder, page 163).

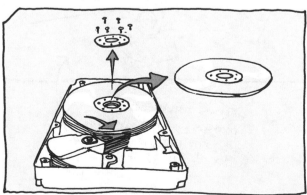

4. PULL OFF THE ACTUATOR ARM

Lift the actuator arm from the hard drive casing. Use scissors to cut the cable connecting the arm and the drive. (It should then be easy to remove the actuator, since there is no longer anything holding it in place.)

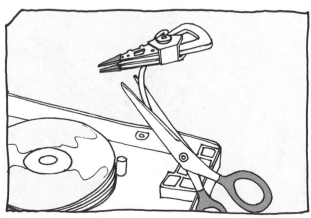

>>>>

5. PULL OUT THE OTHER MAGNET

Remove the other magnet from the hard drive case as shown, then trim off any excess ribbon cables or wire from the actuator arm.

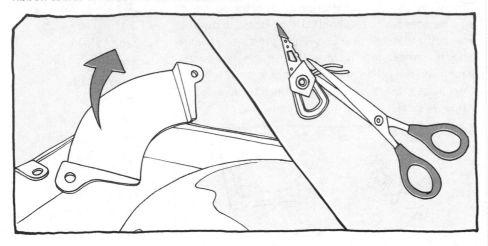

6. REASSEMBLE THE PIECES

To reassemble, touch the magnets together on one side of the assembly, slip in the actuator arm, and then quickly pivot down the other side of the magnet to close the assembly. (The actuator arm should be sandwiched in between the magnets, as it was originally in the hard drive.)

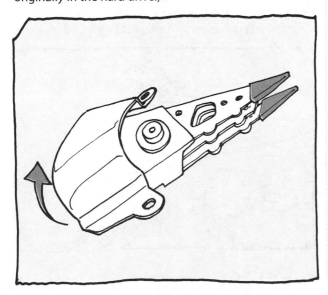

7. THIS IS A STICK-UP

Stick the actuator arm to your refrigerator or other magnetic surface and put it to work holding a photo (or grocery list, or piece of artwork) for you.

This magnetic clip makes a fine addition to the magnetic Memo Board, page 70.

#6 Scanner COMPOST BIN

LEVEL 1: Novice | **TECH TRASH:** Scanner

People have always had a fascination with decomposition. The notion that a living being could break down into the earth has captured the imagination of countless generations. Thanks to advances in scientific methods and instrumentation, we now understand very well how living matter decomposes into soil. Even so, that does not mean we have to stop marveling at and celebrating in the regenerative life process! Use a dead and decaying scanner and add some organic food scraps to create a compost bin that gives new life to your waste. And once you've composted enough potting soil, grow an array of plants and vegetables in the garden to start the process all over again.

MATERIALS

- Broken scanner
- Screwdriver
- Razor blade and/or dental spatula
- Power drill
- Scissors
- Old T-shirt
- Hot-glue gun
- Newspaper
- Spray bottle and water
- Food scraps
- Red worms

MAKE IT

1. EMPTY THE SCANNER

Open up the scanner by removing its screws and pulling it apart. Remove all of the electronics inside.

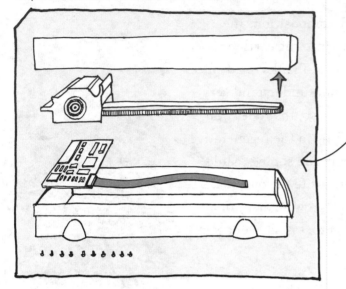

Always save your screws, since you'll likely need them to reassemble.

2. REMOVE THE GLASS

Take the glass out of the scanner bed. This may be as simple as removing a few screws or as complicated as sliding a razor blade or dental spatula between the plastic and the glass to separate the glue. Once the glass is removed, put the scanner case back together.

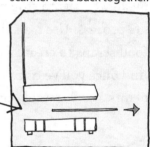

3. DRILL HOLES

Drill about twenty-four ⅛" holes as near to the top of the scanner as possible so that air gets into the compost bin.

4. COVER HOLES

Cut some rectangular scraps from an old T-shirt and glue them to the inside of the scanner over the holes you just drilled so that the worms do not escape. Use hot glue to seal off any holes in the bottom of the scanner—also to keep the worms from wriggling out.

Note: If there are any really large holes in the bottom of the case, you can cover them with plastic panel coverings made from large water bottles. See the Sustainable Vase project Steps 3 through 6 starting on page 78 for a tutorial.

5. BEDDING

Cut a section of newsprint into 1" strips and stir them up into a big, jumbled mess. Place this mess inside the scanner. Dampen the newspaper nest with a spray bottle so that it has the feel of a partially dried out sponge.

6. COMPOST

Load up your compost bin with fruit peels, coffee grounds, produce, and cleaned eggshells. (No meat, eggs, leftovers, or dairy products.) Wash your compost to help get rid of fruit fly eggs and other nasties.

7. ADD WORMS

The moment has come to fill the bin with worms. Cut some more newspaper strips, spread the "bedding" over the worms, and say, "Goodnight, worms. Happy composting."

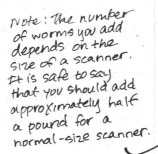

Note: The number of worms you add depends on the size of a scanner. It is safe to say that you should add approximately half a pound for a normal-size scanner.

8. COLLECT

When the container starts to fill with crumbly soil, it is time to collect your compost. First, however, you must separate your worms from the compost. There are a couple of methods for doing this. The most hands-off method is to remove the remainder of the bedding, push all of the compost to one side, and fill the other side with fresh bedding. Start to place food scraps in the new bedding and, over the course of a few weeks, all of the worms should migrate from your compost into the new bedding material. The compost is now ready for collection.

Extra credit: If you are impatient, build a sifting tray to separate the worms from compost.

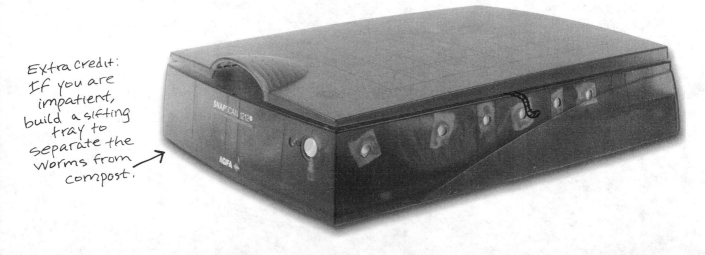

#7 KEYBOARD Lamp

LEVEL 2: Intermediate | **TECH TRASH:** Keyboard

Naked lightbulbs are obscene! Isn't that why lampshades were developed? Okay, okay, maybe it has something to do with diffusing light to make it easier on the eyes. Lamps enable us to benefit from the convenience of a lightbulb without having to stare directly at the light source. These days lamps have come far beyond function to high art. That is what makes lamps so spectacular. Here is a project that takes parts from a nonfunctioning keyboard to make a fully functioning, not to mention incredibly stylish, light source. And it is for this reason that we can never have too many lamps hanging around, especially stylish ones. In the words of weatherman Brick Tamland, "I love lamp."

⚡ SAFETY FIRST

>> **Warning! This project uses wall current. Always consult a licensed electrician when working with wall current and before plugging in your device.**

MATERIALS

- ▸ Keyboard made after 1990
- ▸ Phillips-head screwdriver
- ▸ Hard-drive platter or CD* (A hard-drive platter is ideal. For directions on finding and removing a hard drive platter, please refer to "Actuator Arm Magnetic Picture Display.")
- ▸ Packing tape
- ▸ Small cup
- ▸ Two-pronged power cord (cut from any broken TV, VCR, DVD player, or stereo.)
- ▸ Wire stripper/cutter
- ▸ Light socket with screw terminals
- ▸ Multimeter
- ▸ Hot-glue gun
- ▸ Energy-saving compact fluorescent bulb

MAKE IT

1. OPEN THE KEYBOARD

Open the keyboard by flipping it over and removing the screws. Remove the two sheets with the circuit tracings on them. Set these sheets aside.

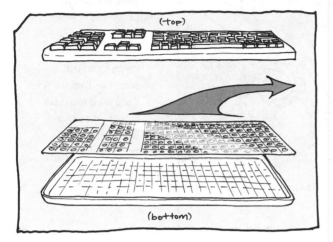

(top)

(bottom)

2. ELEVATE THE PLATTER

Place your hard drive platter (or CD) on top of the cup. Center it over the lip.

3. WRAP IT UP

Wrap one of the plastic sheets lengthwise around the circumference of your platter to create a tube. Tape the seam to hold the plastic in its shape. Set it aside.

You can use the other sheet you set aside to make another lamp.

Fun and Functional
Aside from being more efficient, fluorescent bulbs also produce less heat, which is ideal for a lamp that is attached with hot glue!

4. UNCAP

Remove the socket cap from the light socket assembly by unscrewing it or prying it off.

5. INSERT THE CORD

Pass the power cord through the platter and the socket cap. (If you are using a CD, make sure that the cord passes through toward the reflective side.)

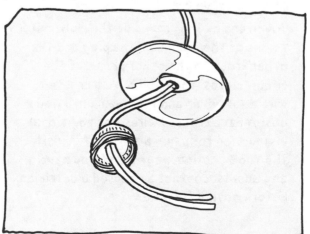

6. KNOT IT!

Split the last 5" of the power cord into two separate insulated wires. Knot the cord ends in an underwriter's knot. Try to leave yourself about 1" of wire on each side to work with. (Be patient, this may take some practice.)

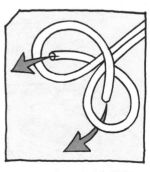

>>>>

7. WIRE THE SOCKET

Strip up to ½" of insulation from the end of each wire. With your fingers or pliers, bend the exposed wire into clockwise loops. Secure each wire to the screw terminal by placing the wire beneath the screw head and tightening the screw.

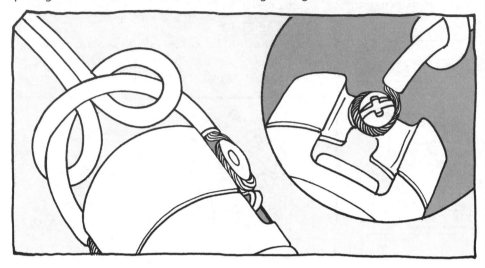

8. REASSEMBLE

Carefully reassemble the socket. Make sure that your wires haven't gotten crossed by testing the continuity of the plug with the multimeter.

Multi-what?

Touch one of the probes of the multimeter to one of the prongs on the plug and the other probe to the other prong. If the circuit closes, then you have a crossed wire. Start over and try again until you do not have crossed wires. For additional instruction for using a multimeter, visit the **Tools** section, page 19. If you have any doubts, contact a licensed electrician before moving forward.

9. GLUE

Position the socket so that it sits flat and hold the platter up like a table. Glue the platter in place so that it is perfectly level.

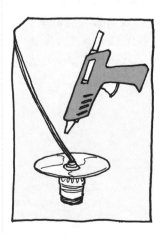

10. MOUNT THE SHADE

Hold the platter in place inside the circuit board tube ⅛"
below the top. Apply hot glue along the seam to fasten it
in place. Hold it steady until it dries.

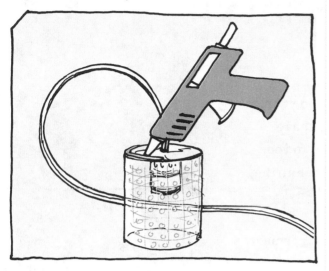

11. SCREW IT

Screw the compact fluorescent bulb into the socket and
hang your lamp somewhere lacking in light and style!

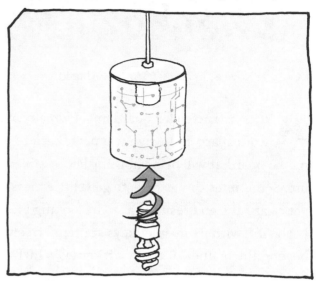

KEYBOARD LAMP

>> variation

* **If you get on a lamp-making kick,
try hanging multiple lamps to create a
chandelierlike effect.**
* **Show your colorful personality:
The standard white bulbs can be replaced
with colored bulbs.**

Be a man
or woman
of letters
with your
nifty new
keyboard
lamp.

#8 MAGNETIC Memo Board
(& Keyboard Magnets)

LEVEL 1: Novice | **TECH TRASH:** Keyboard

A deconstructed keyboard might wreak havoc on your work space, but when you put it back together as a memo board, it will bring nothing but organization to your life. Sometimes things have to get a little messier before they get clean! The end result is a smart-looking Picasso-style keyboard, with its magnet keys scattered freely across it, well beyond the bounds of the traditional QWERTY arrangement.

Mr. Resistor Man Says:
Magnets can erase or damage a number of types of computer storage devices. Keep them away!

MATERIALS
▸ Keyboard
▸ Screwdriver
▸ Scissors
▸ Magnetic tape
▸ Hot-glue gun
▸ Two coarse-thread drywall screws

MAKE IT

1. OPEN THE KEYBOARD

Open up your keyboard case by removing the screws in the bottom of the case.

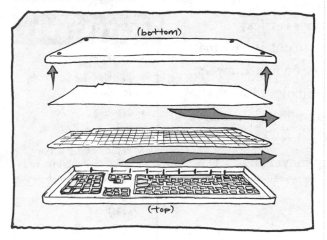

2. POP OFF THE KEYS

Pry off a series of keyboard keys that strike your fancy. Cut the peg in the back of each key so it is just slightly recessed from the back rim of the key.

HELPFUL HINT

You can pop the keys off with your fingers, or if you want to be quick about it, pry them out with a screwdriver.

Don't throw away the other parts! They can be used to make a keyboard lamp (page 66) or a squid-skinned camera case (page 57).

3. PREPARE THE MAGNET

With scissors, cut a piece of magnet that will fit inside the back of the keyboard key and peel off the backing. Then glue the magnet into the key with the magnetic end facing out. Repeat until you have made a handful of key magnets.

4. SCREW TO WALL

Fasten the panel to the wall with two screws. If you are installing it in a wall stud, drill two pilot holes. Otherwise, if you are installing it in drywall, you can just twist it right in.

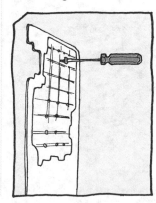

MAGNETIC MEMO BOARD

>> Variation

For more of a "techie" look, place one of the clear plastic circuit boards over the panel before fastening it to the wall.

5. USE IT!

Decorate your magnet board with pictures, papers, and random scraps to your heart's content.

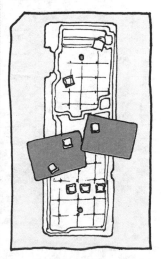

#9 CD-Drive Desk Organizer

LEVEL 1: Novice | **TECH TRASH:** CD-ROM Drive

For most people, consolidation of one's belongings is tricky business. I'm always trying to figure out where to put all my small tchotchkes and knickknacks. There never seem to be enough drawers, compartments, shelves, and interdimensional wormholes to store all of these odds and ends. Wouldn't it be great if you could use your old space-sucking computer to make storage? The answer is yes. Yes, it would be. And that is precisely what you will do by converting a CD drive to a desk organizer.

MATERIALS

- CD-ROM drive
- Phillips-head screwdriver
- Diagonal cutters
- Hot-glue gun
- Power drill with a ⅜" metal bit and an ⅛" plastic bit
- Four ¼" metal washers
- Two C-clamps
- Four #8 by ¾" round-head wood screws (Note: If your desk is less than ¾" thick, do not use ¾" screws. Use screws less than the thickness of your desktop.)
- Ribbon cable

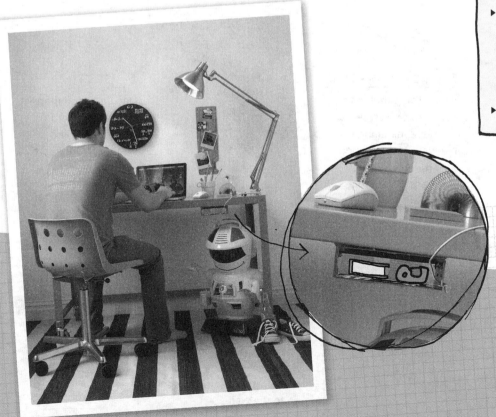

MAKE IT

1. OPEN UP THE CASE

Open up the CD-ROM case by removing the outer screws and prying the case open with your screwdriver. Save the screws for later reassembly.

Keep track of those screws!

2. REMOVE ALL THE GOOD PARTS

Remove everything from inside the case for later use. (After all, a CD-ROM drive is full of useful parts such as motors, gears, and springs. Gather up these parts and set them aside for future projects, such as the Gear Clock, page 56).

3. PREPARE THE FRONT PANEL

Leaving one plastic tab intact (you'll need it in Step 5), cut or break off the rest of the tabs used to fasten the front panel of the drive to the case.

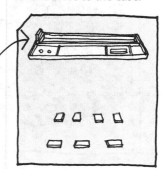

Leave this tab intact.

4. RECONSTRUCT THE FRONT PANEL

Remove the CD tray cover from the sliding CD drawer and glue the CD tray cover back onto the front panel. *Optional:* Glue the audio jack, volume knob, buttons, and LEDs back in place.

Note: If you are having trouble removing the volume knob from the circuit board, you can convincingly use a plastic gear in its place.

5. DRILL A SMALL HOLE

Drill an ⅛" hole through the plastic side tab you left intact on the front panel in Step 3.

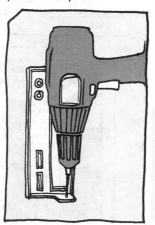

6. DRILL SOME BIG HOLES

Clamp down the top half of the drive's metal case and mark and drill four ⅜" holes positioned in a large rectangle on the case. These will be used for mounting the drive to the underside of the desk.

>> >>

7. BOLT TO DESK

Place a washer over each screw hole between the case and the desk. Fasten the top section of the case to the underside of the desk using wood screws.

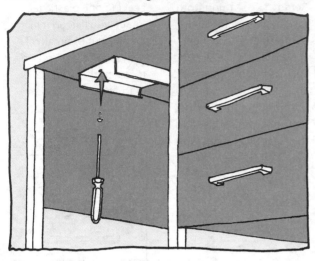

8. CONNECT BOTTOM

Reattach the bottom part of the case using the screws you set aside in Step 1.

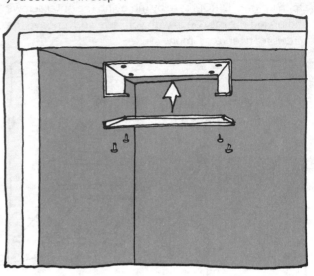

CD-DRIVE ORGANIZER

>> variation

Install the organizer on a wall to make a takeout menu holder or a makeshift mailbox.

9. INSTALL THE FRONT PANEL

Pass one end of the ribbon cable through a hole in the side of the case and knot it so it won't slip out. Pass the opposite end of the ribbon cable through the hole you drilled in the front panel and knot it. Once complete, clip on the front panel back to the case.

10. FILL WITH STUFF

Fill your cubby with all of the wonderful treasure that litters your desk.

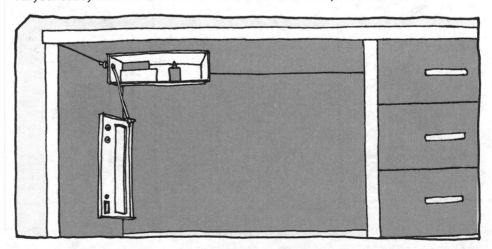

#10 DRIVE Bookends

LEVEL 1: Novice | **TECH TRASH:** Old Computer

You can pull a pile of useless little drives out of most old computers. There will usually be a floppy drive, maybe a Zip drive, a hard drive or two, and, perhaps, some other inexplicable mystery drive. Aside from containing a couple of very useful parts, these drives are more or less a pile of junk only suitable for use as paperweights. So, what is a person to do with a stack of paperweights? Hold up a stack of books!

MATERIALS

- Old computer
- Phillips-head screwdriver
- Book (or other flat-sided object)
- Mixing stick
- Epoxy

Mr. Resistor Man Says:
When invented in 1956, the first magnetic hard disk drive was the size of a refrigerator and stored 4.4 megabytes of data. That is about enough space to store one MP3.

MAKE IT

1. REMOVE

Unscrew and pry off the side of your computer. Remove all of the old disk drives—there should be three to five—from inside.

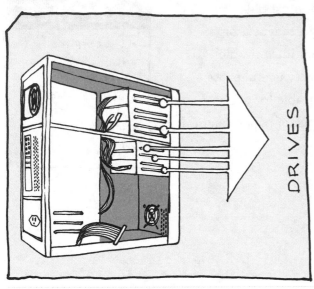

2. STACK

Stack the drives until you find an interesting configuration you like.

3. SQUARE

Take a book (preferably hardcover) and press it against one side of the bookend pile to make sure that it will hold the book upright while maintaining your artistic configuration.

4. STICK

With a mixing stick, stir together the two-part epoxy. Then spread epoxy between the layers of drives and press them together.

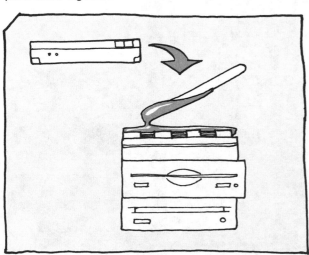

Hint: Use the packaging the epoxy came in as your mixing surface.

5. DRY

Let dry for as long as directed on the epoxy packaging. Read a book while you're waiting—then return it to the shelf propped up in techie style!

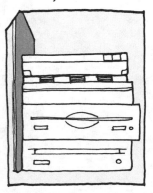

#11 Sustainable Vase

LEVEL 2: Intermediate | **TECH TRASH:** External Disc Drive

One of the greatest ways to liven up any home is with a bouquet of freshly cut flowers (which, sadly, are slowly withering for your enjoyment). Of course, if you don't enjoy watching flowers die, preserve them a little while longer by sticking them in some water. The same thing could be said for external hard drives . . . well . . . sort of. The big difference is rather than sticking the drive in water to extend its life, you stick the water in the drive. Once the water's in the drive, you have a lovely vase to preserve your flowers with!

MATERIALS

- External disc drive
- Screwdriver
- 2-gallon plastic water jug
- Scissors
- Rubber gloves
- Silicon aquarium glue
- Hot-glue gun
- Craft knife

Mr. Resistor Man Says:
The IEEE is the leading professional technology association and is responsible for establishing technology standards, such as "IEEE 1394," better known to most by the trademarked name, FireWire.

MAKE IT

1. TAKE APART THE DRIVE

Remove the outer casing of the drive by unscrewing it or otherwise prying it open. Remove the innards (the electronic components) and set them aside for use in other projects.

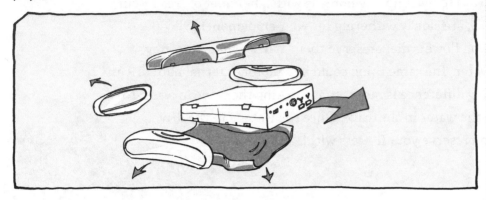

⚡ **SAFETY FIRST**

>> This project requires a well-ventilated space. Always carefully read the safety labels on the packaging before working with any type of glue.

2. REASSEMBLE

Reassemble the empty plastic case, clipping or screwing it back together. Identify where the seams are and which end of the drive will be the bottom.

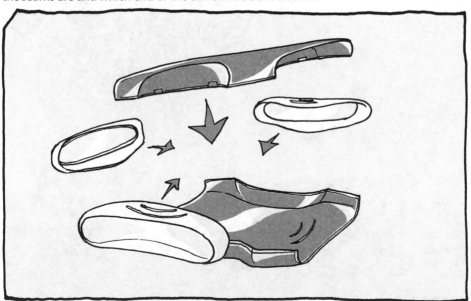

3. PATCHES

Locate any holes through which water could leak. Take an empty water jug and use scissors to cut rectangles of plastic big enough to cover large holes in the casing. *Note:* Don't worry about patching screw holes which can be filled later with aquarium glue.

4. GLUE IT

Put on your rubber gloves and squeeze a generous amount of aquarium glue around the edges of the patches, one patch at a time, and press them in place so that the holes are covered. Stack something heavy to weigh down your patchwork to ensure a good seal with the case.

5. SEAL IT

Apply a ¼"-wide bead of glue along all the inside seams of the container to further seal it.

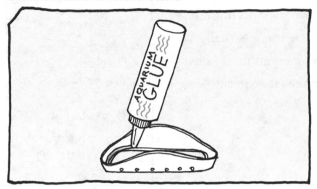

6. WAIT AND VENTILATE

The glue can take one or two days to fully set. Some forms of aquarium glue produce nasty fumes. Leave your vase outside to dry or near an open window with a fan pointing outside.

7. CLEAN IT UP

With a craft knife, carefully trim off any excess glue that might have hardened on the outside of the vase, but be careful not to trim away too much, or it may start to leak.

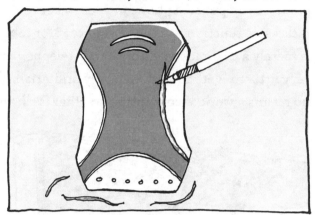

8. FILL WITH WATER

After the glue has dried completely, test your work by filling the vase with water. Keep it over a sink in case it does leak. If it's not leaking, you've done a great job sealing it. Fill it with flowers and enjoy! If it's leaking a bit, see where the water is coming from and how large the leak is, then proceed to the troubleshooting section.

Troubleshooting

FIX THE LEAK

First empty the vase. If the leak is very small, you may be able to fix it with a small dab from a hot-glue gun. If it is larger, apply more aquarium glue and wait another day or two.

If you can't fix that leak no matter what you try, fret not. Drill a few holes in the bottom and make a planter!

#12 "Please Hold" Phone Rack

LEVEL 1: Novice | **TECH TRASH:** Old Phone

Let us clear up any confusion here and now. A phone rack is not a rack for holding phones. It will not hold rotary phones, push-button phones, or even wireless phones. It will maybe hold a cell phone if you put said phone in your jacket first. For you see, a phone rack's sole function is to place your coat (or robe, or towel) on hold. It is merely a coat rack made out of old telephone parts. There is really very little to "get" about this highly utilitarian storage system. It is by no means a work of conceptual art. That is what makes it so great.

MATERIALS

- Old phone
- Modular telephone jack
- Phillips-head screwdriver
- Diagonal cutters
- Pliers
- Two ³⁄₈" nuts
- ³⁄₈" by 24" threaded rod
- Two 1" L brackets
- Two 1" self-tapping screws
- Zip tie
- Black marker
- Hacksaw
- Spiral phone cord
- Power drill
- Two 1¼" wood screws

Mr. Resistor Man Says:
Unlike cell phones, rotary phones were designed and built to continue working for decades (and still work on many phone systems).

MAKE IT

1. OPEN

Pop and lift the cover off the modular telephone jack, unscrew the terminals inside, and clip and remove all of the wires.

2. THREAD

Insert a ⅜" nut directly behind the opening at the front of the phone jack and thread the rod through both the jack and the nut, as shown. Hold the nut in place and continue passing the rod through until it nearly reaches the back of the jack. Place another nut at the back of the modular telephone jack and thread the rod through the second nut.

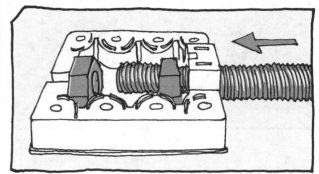

3. BRACKETS

The L brackets should be mounted to the modular telephone jack so that the perpendicular part of the bracket is aligned to the top of the jack (the end with no threaded rod protruding). Using 1" self-tapping screws, attach two L brackets to the back of the jack, aligning the holes in each L bracket with the predrilled holes in the telephone jack casing. Ensure that they are firmly and evenly secured in place.

4. FASTEN

There should be at least two mounting holes in the back of the jack case. Attach a zip tie through these holes so that they pass between the nuts and hold the rod securely against the back of the modular telephone jack. Tighten and snip the end.

5. MARK

Rest the protruding ends of your L brackets on the top of a door (letting the rod lie against the back of it) and make a marking on the rod 1" below where you would like the phone handle to be mounted. Use the hacksaw to cut the rod to size at the marking.

>>>>

6. CLOSE

Slide the cover of the modular telephone jack (from Step 1) up the length of the rod and fasten it back in place over the jack.

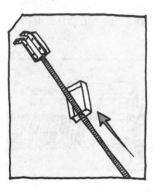

7. COVER

Slide the end of the rod through the spiral phone cord until the rod is hidden completely. Trim off any excess coil that extends beyond the end of the rod.

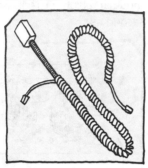

8. DRILL

Drill a shallow ⅜" hole that is centered upon the side of your handset. The hole should barely go through the side of the case and should be just large enough that you can start twisting the threaded rod into the handset.

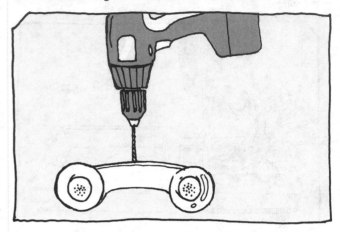

9. THREAD

Cut away about 1" of the coiled phone cord and start to thread the rod onto the phone handset. Since the space between the inner walls of the handset's handle is presumably narrower than ⅜", the action of threading the rod into the handset should provide enough tension to hold the handset in place.

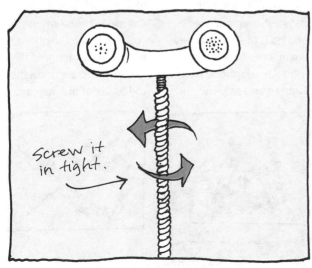

Screw it in tight.

10. MOUNT

Make two marks through the holes in the L brackets on the top of your door. Drill ¹⁄₁₆" pilot holes into the door at those marks. To mount your phone rack, insert wood screws through the L brackets and into the drilled holes.

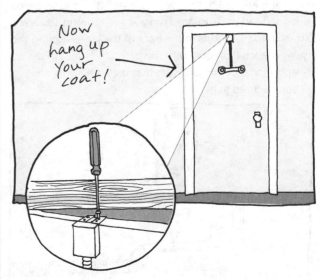

Now hang up your coat!

#13 Phone Safe

LEVEL 2: Intermediate | **TECH TRASH**: Telephone Handset

Hollowed-out books are so 19th-century, and false-bottomed cans are so 20th. If you want to keep unsavory characters out of your business today, you're going to need to get a lot more creative. No more undersides of drawers or high furniture ledges—it's too obvious. It's time to get crafty. Really think like a thief. A wireless telephone handset without its base, for instance, has little to no resale value. It is so ordinary that no one will suspect it is hiding any secrets, and since it has little resale value, the likelihood of your casual burglar snatching it is low. Let's just hope that those nefarious creatures of ill intent aren't reading this book. . . .

MATERIALS

- Telephone handset
- Phillips-head screwdriver
- Diagonal cutters
- 2-gallon water jug (or gallon milk carton)
- Scissors or craft knife
- Hot-glue gun
- Quick-setting epoxy
- Q-tips (optional)
- Two small magnets

MAKE IT

1. OPEN IT

Unscrew and open the handset casing. Set aside the screws for later reassembly. Get rid of the base unit. (You don't want that hanging around, or someone might see resale value and grab the entire set.)

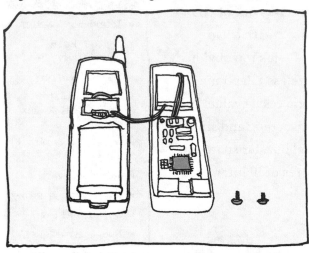

2. GUT IT

Remove the electronic circuitry, wiring, and speakers. You can leave in the button pad and screen. Use your diagonal cutters to cut out any excess plastic (including the sockets that the screws fit into) that isn't used to keep the case snapped shut. Also, remove any components from the circuit board that are visible from outside the phone, such as the headset jack.

3. JUG

Cut out a piece of plastic from the water jug that is just large enough to cover the backside of the button pad and screen. Then use your hot-glue gun to glue the plastic piece in place.

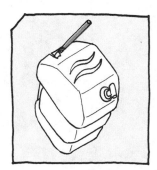

4. MAGNETISM

Follow the packaging instructions to mix up two-part epoxy. Glue the magnets over the sockets, placing them so they won't interfere with the case being closed. *Note:* Temporarily remove any metal pieces from the case that may be near enough to the magnets to attract them. (Once the epoxy is fully set, put the metal pieces back.)

Tip: cut off the fuzzy end of a Q-tip to make a good stirring stick for mixing two-part epoxy.

5. MORE GLUING

After the epoxy has dried, hot-glue the rest of the parts (such as the headset jack) to the case so that the phone looks as inconspicuous as possible.

Add all the parts that will make the phone look most realistic.

6. MORE TRIMMING

Use your diagonal cutters to trim away most of the threading on the screws you set aside earlier so that only ⅛" to ¼" is left.

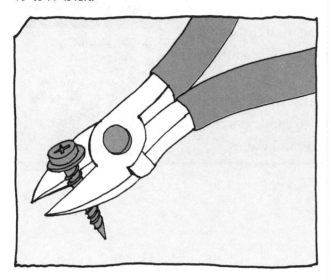

7. EVEN MORE GLUING

Close the two sides of the case and apply a small amount of hot glue inside the screw socket. Tightly press the screw into the socket and hold it until the glue dries. Once dry, scrape off any visible glue. The magnets will now pull on the screws to keep the case closed. *Optional:* To make a very tight seal, consider putting magnets on both sides of the case.

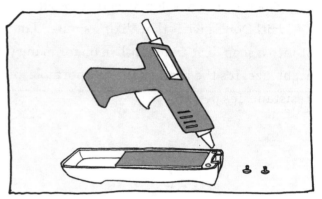

Tip:
Don't forget to put the battery back into the phone! This will add enough weight to make it believable as an unaltered telephone.

8. HIDE YOUR VALUABLES

You are now able to safely stash all of your cold hard cash and precious little treasures.

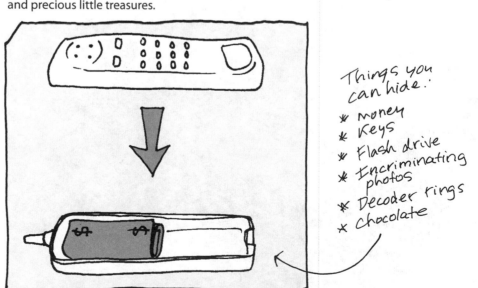

Things you can hide:
* Money
* Keys
* Flash drive
* Incriminating photos
* Decoder rings
* Chocolate

#14 Walkman Soap Dish

LEVEL 1: Novice | **TECH TRASH:** Walkman

It has been with you through thick and thin. You've shared moments of joy, heartbreak, defeat, and triumph. And through it all, it has remained by your side. Now, after many years and countless mix tapes, your old walkman is truly broken.

Wait! Don't give it that Viking's funeral quite yet. Instead, turn it into a soap dish and let it live on for many years to come. It's the right size. It's the right shape. It's portable, durable, and water-resistant. It's perfect!

MATERIALS

- Old and/or broken walkman (preferably of the "sports"/water-resistant variety)
- Set of small screwdrivers
- Hot-glue gun

Mr. Resistor Man Says:
The walkman was invented at the behest of Masaru Ibuka, the honorary chairman of Sony, who was tired of carrying a bulky tape recorder with him on long plane trips.

MAKE IT

1. REMOVE THE ELECTRONICS

Using your screwdriver set, unscrew and remove all of the electronics, mechanical parts, and any extra metal plating from the inside of the case.

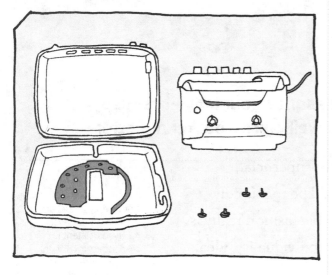

2. GLUE IN THE BUTTONS

Glue in place the buttons for play, stop, rewind, and fast-forward, as well as any other buttons that will fall off if they are not attached to the case. *Note:* For any switch that is held in place with a metal screw (such as the volume knob), remove the screw and glue the switch or knob in place.

3. PREPARE FOR USE

Once it is dry, rinse the inside of the case and put your soap inside.

After years of "walking," your walkman deserves a little R&R.

CHAPTER 3

Fashionable Technology

> Go geek chic with 12 projects to accessorize your look—because there's so much more to style than a well-placed pocket protector.

S
ome may write off fashion as frivolous, superficial, mindless, or petty, but to do that would be to stupidly discount how extraordinarily remarkable fashion really is. Through fashion we can instantly express our individuality, ideals, and relation to the social whole.

It is, therefore, fair to say that with a single glance we can look at a person and tell their story. It's also true that the story we tell may not always be correct (the old "judging a book by its cover" doesn't always work), but a large portion of the time, it will provide enough insight to make the snap judgment worthwhile. In terms of making generalities about a person and the culture to which they belong, fashion is a wonderful tool.

Mr. Resistor Man Says:
Nerd alert! Samuel Clemens (Mark Twain) received a patent for suspenders in1871. In 1894, inventer David Roth was issued a patent for metal-clasped suspenders we're familiar with today.

PROJECTS

Ribbon cable hair clip, p. 102

In light of this, how we choose to (or not to) present ourselves to the world is very important. Depending on what we wear (or don't wear), others can infer a great deal about who we are and what we stand for. We are not just throwing on clothes in the morning; we're putting together a visual representation of ourselves. In fact, I am pretty certain that unlike myself, some people spend a lot of time in the morning carefully preparing their outfit. This deliberateness in styling makes outwardly visible an astute attention to detail and calculated worldview that can be used to easily distinguish this person from a laid-back slob like me.

Before you spend too little or too much time getting ready tomorrow morning, ask yourself, "What story do I want to tell?" Perhaps you want to convey your embrace of the digital age by wearing the carcasses of dead computers around your neck. Or, maybe,

Cable corsage, p. 109

RAM Money clip, p. 116

Hearmuffs, p. 107

8-Bit Belt Buckle, p. 112

Keyboard Buttons, p. 104

you would like to announce to the world that in this troubled time reuse is cool and recycling is sexy. Or, perhaps, you want people to think you are a big dork by letting them know that these high-tech electronic trinkets that can be easily recognized but rarely defined are beautiful. Then again, maybe you have another reason altogether. Maybe you're just trying to save money and be stylish.

In some capacity the following chapter can help you with all of these goals. There are a number of fashion ideas to inspire you in your daily life. Remember that fashion is a series of contrasting relationships between the things you wear and how you wear them. You can make as loud or as subtle a statement as you would like. It really comes down to how you would like to express yourself. Identify a few projects that you like and experiment until you find what combination suits you.

#15 QWER-Tie

LEVEL 1: Novice | **TECH TRASH:** Keyboard

Nothing quite says "sexy" like keyboard keys. That is, if your idea of sexy is the ability to type 120 wpm, pwn at *WoW*, and set up an effective firewall in XP. If it is, then this little fashion accessory will greatly aid in fluffing your Linux-loving penguin feathers and attracting the ideal mate. All you need to do is to secure the QWER-Tie ponytail holder in your hair, go to an MMORPG meet-up, and chase after the first dork that backs away from you.

MATERIALS

- Keyboard
- Thin hair rubber band(s)
- Screwdriver
- Diagonal cutters
- Power drill with a ⅛" plastic bit
- Hot-glue gun

Mr. Resistor Man Says:
The QWERTY keyboard layout has been in existence since the early 1870s, almost directly coinciding with the invention of the typewriter.

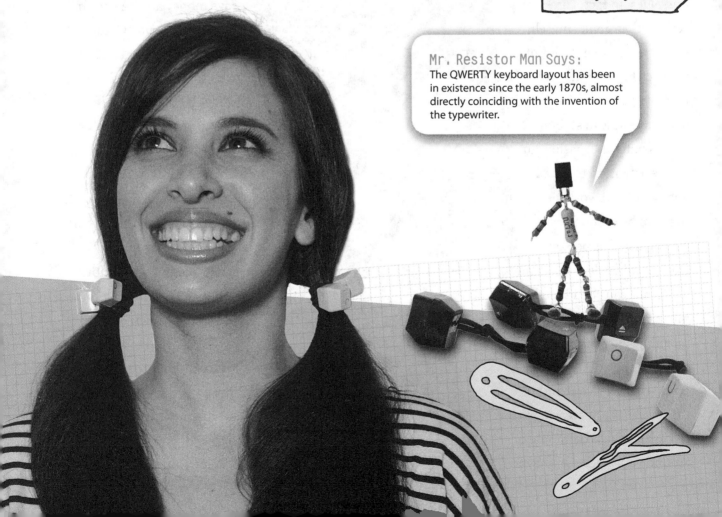

MAKE IT

1. TIE A KNOT

Tie the rubber band into a single overhand knot that splits it in half. Pull the knot very tight.

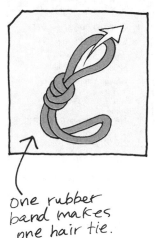

one rubber band makes one hair tie.

2. REMOVE SOME KEYS

Using a screwdriver, pry off a handful of keyboard keys. Select the four you like best. (Some ideas: Go completely random, or try N, S, E, W for directionals, spell L-O-V-E or other four-letter words with nonrepeating letters, maybe your name has four letters, or perhaps you just like punctuation combinations ?!&?)

3. TRIM THE FAT

Select two keys and use the diagonal cutters to trim out all of the excess plastic from inside until they are hollow. *Note:* These keys will appear opposite one another on the rubber band.

4. DRILL

Take the other two keys that were not hollowed out and, one at a time, mark and drill a centered ⅛" hole into the side of each.

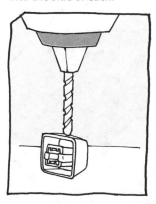

5. PASS IT THROUGH

Pass one end of the rubber band through the hole in one key and hook it around the piece of plastic in the center. Pull on the rubber band to secure it tightly in place. Repeat on the other side of the rubber band with the other key.

6. GLUE IT TOGETHER

Squeeze glue into the back of one of the keys secured to the rubber band. Line up the edges of the hollowed-out keys against it. They should lay flat against each other to form a single bead. Hold the keys in place until they dry. Repeat on the remaining two keys to complete the other side of the rubber band.

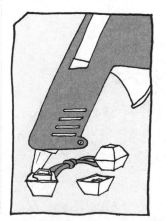

QWER-TIE

>> Variation

★ Add some variety to your life, and try mixing and matching keys from different keyboards—black keys with gray keys, for instance, for a checkerboard effect.

#16 CAPACITOR Earrings

LEVEL 1: Novice | **TECH TRASH:** Circuit Board with Capacitors

One of the most basic and readily available electronic components is a capacitor. Capacitors store a small amount of electricity until they are fully charged and then release the charge all at once back into the circuit. This property makes capacitors essential for a number of high-tech applications. When you throw away an electronic device, chances are you are discarding dozens of perfectly good capacitors that can still be used for many things from the practical to the whimsical and fun. Capacitor earrings definitely fall into the latter category. However, if you're ever stranded on a desert island and in dire need of a 10µF capacitor to increase the broadcast range of your coconut radio, you will be glad you made these, you resourceful techie fashionista, you.

MATERIALS

- Circuit board with capacitors on it (found in scanners, printers, computer power supplies, VCRs)
- Wire cutters
- Needle-nose pliers
- Solid wire
- Wire strippers
- Two earring hooks (french hooks are ideal)
- Soldering iron (optional)
- Lead-free silver solder (optional)

Mr. Resistor Man Says:
Due to a lack of standardization in the manufacturing process, electrolytic capacitors can contain a wide array of electrolytes from the benign to the toxic and even to the hallucinogenic.

HANDMADE BY _____

DIY

MAKE IT

1. IDENTIFY AND GATHER CAPACITORS

Capacitors either look like colorful tubes rising up from the surface of the circuit board or flat discs with two little metal legs (to learn more about capacitors, see page 10). Find one that tickles your fancy, and carefully snip it off, leaving as much lead wire as possible still attached. Repeat to collect six to ten capacitors.

FYI: The circuit board these capacitors came from was inside an old scanner.

2. BEND THE LEADS

With your pliers, bend the two lead wires on each capacitor to form a loop.

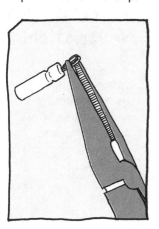

3. STRIP AND TRIM

Take a piece of solid wire and strip off the insulation. Trim the stripped part of the wire so it's about 2" long.

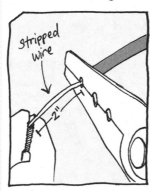

stripped wire

2"

4. BEND THE WIRE

Bend a small U-shaped loop at one end of the wire and slip this loop through the bent leads on the capacitor. Then bend the U-shaped loop closed with your pliers.

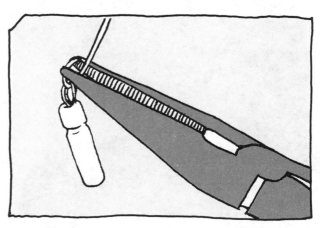

continued >>

5. HOOK IT ON

Slip your wire through the loop at the bottom of the earring hook and bend it so that it folds with the wire in half.

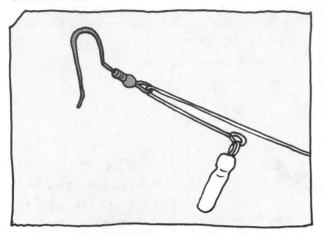

6. TRIM OFF THE EXCESS

Cut off any wire that extends past the capacitor. It should appear to be folded neatly in half.

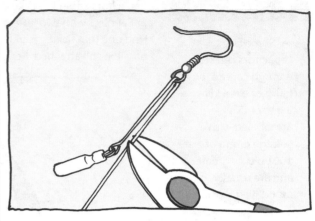

7. REPEAT

Repeat Steps 3 to 6 until you have an even number of capacitors dangling from each earring hook. Try them on or give them as a gift. *Optional:* For a more permanent connection, you can apply a small amount of lead-free solder to close all of the loops that you have made. (See page 34 for a soldering refresher course.)

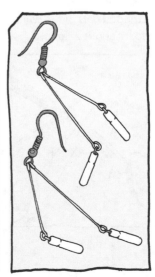

CAPACITOR EARRINGS

>> variation

* To liven up your earrings, try incorporating resistors, transistors, and other small electronic components.

#17 ROCKING Remote cuff

LEVEL 1: Novice | **TECH TRASH:** Old Remote Control

There is nothing punk rockers like more than spending an evening at home, drinking a nice cold beverage and watching all of their favorite shows on TV. Well, actually, there is perhaps one thing they like more: a strict adherence to fashion mores. Any real punk rocker can tell you how important it is to wear the proper uniform. It is only natural that one should want to combine punk rock's love for television with its propensity toward visual uniformity.

In other words, there is nothing more punk rock than wearing the all-too-predictable spiked bracelet made from the gummy material inside a remote control. It is the embodiment of the two things punks love the most: television and conformity. Oh, actually, the three things punk rockers like the most. Did I forget to mention? Punk rockers also like irony.

MATERIALS
- Old remote control
- Screwdriver
- Needle or pushpin
- Eyelets
- Eyelet pliers
- Stranded wire

Mr. Resistor Man Says:
The first known patent for a remote control was filed by the infamous inventor Nikola Tesla in 1898 (U.S. Patent 613809).

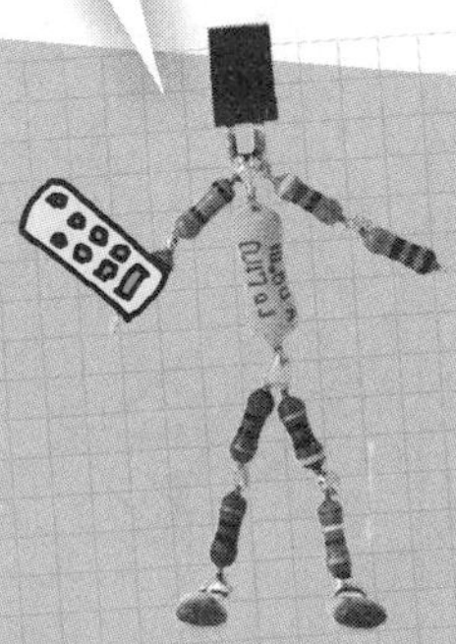

MAKE IT

1. OPEN THE REMOTE

Remote controls are designed to take a beating, so they are difficult to open. After removing the screws, you may need to forcibly pry the case open along its seam with a screwdriver.

2. REMOVE THE RUBBER

Remove the rubbery button pad from inside the remote. Set aside the circuit board for other projects (such as the RAM Money Clip, page 116).

3. POKE HOLES

Use a needle or pushpin to poke two holes, large enough to insert small eyelets through, in each end of the button pad, as shown.

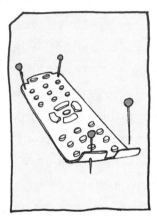

4. INSTALL EYELETS

Push eyelets through the holes from the top to the bottom (so the head of each eyelet appears on the same side as the remote buttons).

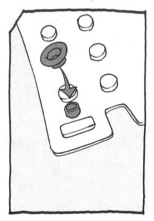

5. CLAMP EYELETS

Put the finished end facedown and clamp the eyelet shut with your eyelet pliers.

6. FLIP AND REPEAT

Turn the button pad and repeat Steps 4 and 5.

7. LACE IT UP

Pass each of the wire ends through corner eyelets on opposite ends of the cuff. Pull tight, then cross the ends over the wrist and thread them up through the remaining two holes.

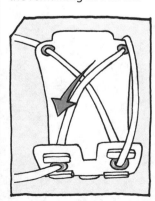

8. TIE IT UP

Tie the ends of the wire in a double knot to hold the bracelet in place.

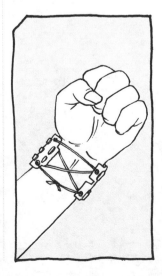

#18 Cascading Pendant

LEVEL 1: Novice | **TECH TRASH:** Old Computer

P eople have been wearing jewelry for far longer than there have been computers to make it out of. To decorate ourselves using resources on hand is something that is inherently human—it helps us define who we are as individuals and assimilate ourselves into the social order. It is only natural, then, in this electronic age for us to bedeck ourselves with the trappings of integrated circuitry. Not only are we affirming our attunement to the world at hand by ornamenting ourselves in recognizable symbols of our age, but we are expressing our individuality by repurposing these once-functional components as a decorative extension of ourselves.

MATERIALS

- Old computer
- Phillips-head screwdriver
- Scissors
- Diagonal cutters
- Needle-nose pliers
- Jewelry wire (copper or brass to match the beads)*
  *Choose a gauge of wire that is strong enough to support the weight of the beads. Larger beads typically require 16 to 20 gauge, while 22 to 24 gauge is sufficient for the smaller-size pieces.
- Necklace cord or chain (about 36" to 42" long)

MAKE IT

1. OPEN YOUR COMPUTER

Remove the screws in the back of the computer and slide the panel off to open up your computer case.

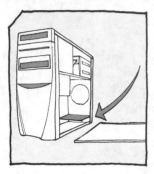

2. TOROIDAL TIME

Look to the computer's motherboard to locate ¾" to 1" in diameter toroidal coils. Snip them off the circuit boards with your diagonal cutters.

3. ALIGN YOUR DESIGN

Arrange the toroidal coils or "beads" flat on your work surface until you find a pendant design that strikes your fancy.

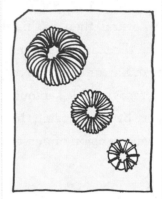

4. IT'S THE TOPS

Take the bead designated as the top or main bead (the one that will later be closest to the necklace chain) and use the needle-nose pliers to bend any loose wire neatly around itself to "finish" your bead (but leave some room to thread some wire underneath later for attaching the pendant chain). This will tuck any pesky, protruding, or sharp metal edges safely away from your skin. Repeat on the remaining beads, mimicking the already existing pattern of wire on the bead.

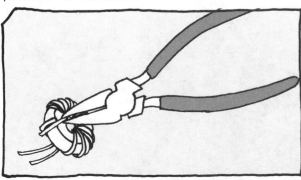

5. GET IT TOGETHER

Take a 6" piece of jewelry wire and thread it through the center of your top bead. Loop it around and through the hole a few times, so that you are left with about 2" of jewelry wire off each side.

⚡ SAFETY FIRST

>> There are normally many large toroidal coils (between .25" and 1.5" in diameter) stored with your computer's power supply. Should you decide you want to remove these, be extraordinarily careful, because the large capacitors inside the power supply can store enough voltage to kill you. If you are unsure about this, stay away from them. For more information, see the Safety section, page 40.

6. CONNECT

Tighten the wire and wrap the ends around the next bead in the sequence at least one full turn.

7. SNIP IT

Trim any jewelry wire that is protruding from the bead with the diagonal cutters, being sure to leave just enough to wrap it neatly around the bead so that you can tidy up all loose ends with pliers.

8. REPEAT

Repeat Steps 5 through 8, alternately threading, wrapping, and tightening the wire around your beads until you are either out of beads, tired of adding beads, or have created such a huge monstrosity that your neck will no longer be able to support it!

Standard DIY means do-it-yourself. In jewelry-making DIY means design-it-yourself.

9. GET LOOPY

Thread a small piece of jewelry wire through the loop that you made (in Step 5) on the top bead. With your pliers, fashion this small piece of wire into a ring or loop. This loop will be used to hang your pendant from your necklace cord.

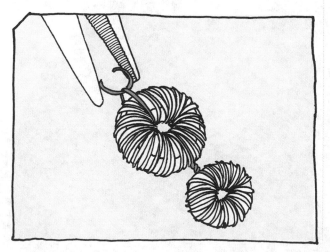

10. KNOT IT

String your necklace cord through the loop from Step 9. Tie a knot around the pendant loop to hold it in place.

11. HANG 'EM HIGH

Pull both ends of the cord around your neck and fasten them in back.

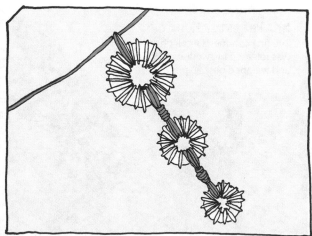

#19 Cell Phone Assassin Necklace

LEVEL 1: Novice | **TECH TRASH:** Broken Cell Phone

In middle school, my history teacher told the class a story about his tour in Vietnam: One night he felt something strange tickle his nose. He brushed it away, and went back to bed. He later awoke, surrounded by men wearing necklaces of human ears. Alarmed at first, he learned that these were mercenaries hired by the CIA to kill Viet Cong in their sleep. They snuck into his bunker to make sure soldiers in his platoon were wearing dog tags. If they weren't, throats were slit and ears added to necklaces. You see, these freelancers were paid on commission, and more ears meant more pay.

The story stuck with me. So, when it came time to transform a dead cell phone, my first thought was to pull out the smart cards and wear them around my neck like trophies. Nothing says "natural born killer" like a chain of smart cards.

MATERIALS

- Broken cell phone (that still has a smart card)
- Small piece of scrap wood
- C-clamp
- Ball chain
- 4 mm silver jump ring
- Needle-nose pliers
- Power drill with 1/16" drill bit

Mr. Resistor Man Says:
The first commercial cell phone was released in Sweden in 1956 and weighed only 90 pounds.

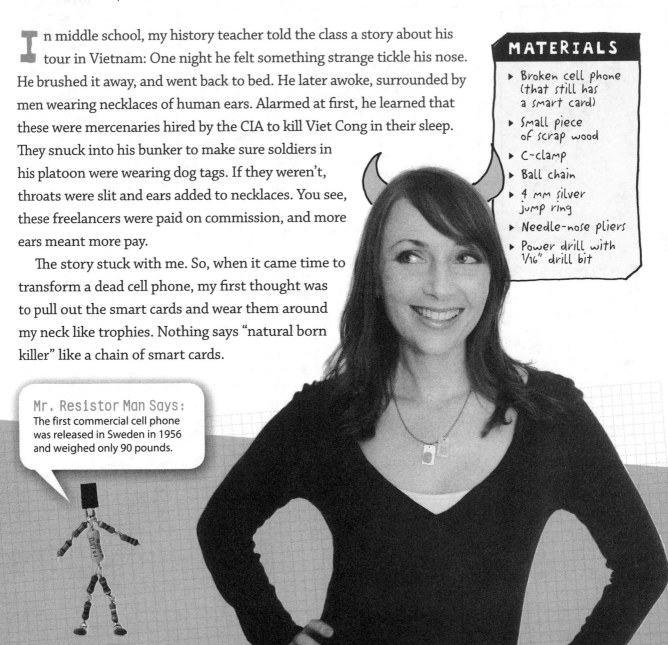

MAKE IT

1. CUT OUT THE BRAIN

Pop open the back of your phone and remove the smart card. If it has one, it should be underneath the battery.

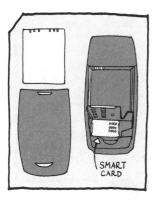

SMART CARD

2. PIN IT DOWN

Place the smart card, shiny side up, on top of the piece of scrap wood at the edge of your work table. Clamp it down gently (so the card doesn't break) with the foot of the clamp on the shiny part.

Be smart! Keep fingers away from the drill.

3. DRILL THE HOLE

Drill a tiny, centered hole about ⅛" in from the edge of the smart card. (The hole should be big enough to fit the jump ring through.)

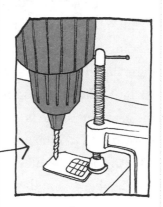

4. STRING IT UP

Pry open the jump ring and insert it through the hole in the smart card. Then place the chain inside the jump ring also.

5. CLOSE AND START COUNTING

Pinch the jump ring closed with your needle-nose pliers. Wear proudly! (And every time you "kill" another phone, add a new smart card to your necklace and rejoice in the wastefulness of planned obsolescence.)

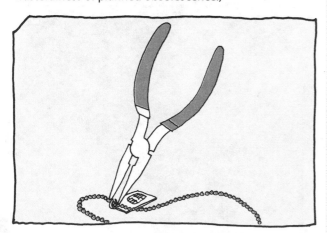

#20 Ribbon CABLE Hair clip

LEVEL 1: Novice | **TECH TRASH:** Ribbon Cable

As drab and gray as most computers are on the outside, they're actually rather colorful on the inside. There are colorful circuit boards covered in capacitors and resistors and, perhaps most exciting, there are ribbon cables. A ribbon cable, as the name would imply, is a result of careful crossbreeding of ribbons and cables. Ribbon cables are made up of many tiny insulated cables that run side by side to form bendable and often brightly hued plastic ribbon. Granted, some of this ribbon cable is gray, but some is rainbow-striped, and some has the same colors you'd find in bins of hard mint candies at a country store—you know, with the white and color swirls?

Now, ribbon cable isn't as flexible as standard ribbon, so don't go tying it right in your hair, but when you attach it to a metal barrette backer, it couldn't make a better hair accessory.

Mr. Resistor Man Says:
When manufacturers first came out with the rainbow ribbon cable (which borrows its color coding from resistors), it was affectionately called "hippie cable" by users.

MAKE IT

1. TRIM THE EDGE

Measure the length of the back (flat part) of the barrette and add ½" to get x. Trim the ribbon cable to the length of x.

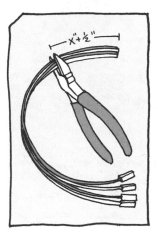

2. GLUE IT DOWN

Squeeze a thin line of hot glue along the top of the barrette, and center and press the cable down on top, leaving the trimmed ends sticking out about ¼" on each end.

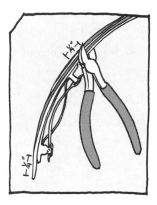

3. GLUE DOWN THE EDGE

Turn the barrette over and squeeze a dab of glue on each end of the cable. Fold the ends over the edges of the barrette and firmly hold in place until the glue sets.

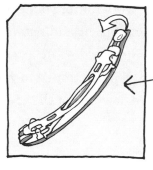

Press the ribbon cable firmly.

⚡ SAFETY FIRST

>> **Be careful not to get hot glue on your fingers. It burns!**

4. PICK OFF THE GLUE

Remove any excess dried glue around the edges with your fingernails.

5. PUT IT IN YOUR HAIR

Clip it in your hair. If you don't have any hair, you can put it in your friend's hair. If you don't have any friends, you can get a wig.

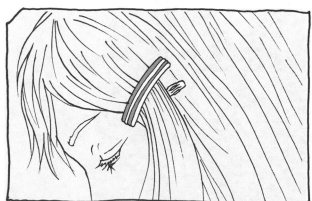

#21 KEYBOARD Buttons

LEVEL 2: Intermediate | **TECH TRASH**: Dead Laptop Computer

ong gone are the barbaric days of the 20th century, when geeks were preyed upon by meatheads intent on dishing out knuckle sandwiches and wedgies. Today, geeks rule the world!

It is time to stop hiding behind your computer and to march heroically into the outside world (after your eyes adjust to natural light, of course). My fellow brothers and sisters, it is now time to stand up for yourself and say to the world, "I am a geek and I am proud!" The time is ripe to wear your geekiness pinned to your chest or displayed proudly on your sleeve. The dawning of the age of keyboard shirt buttons has come!

MATERIALS

- Dead laptop computer
- Screwdriver
- Shirt with buttons
- Scissors
- Gaffer's or masking tape
- Scrap wood for drilling
- Power drill with a 1/16" drill bit
- Needle and thread

Mr. Resistor Man Says:
In 1983, the computer became the first object to receive the distinction of "Person of the Year" by *Time* magazine.

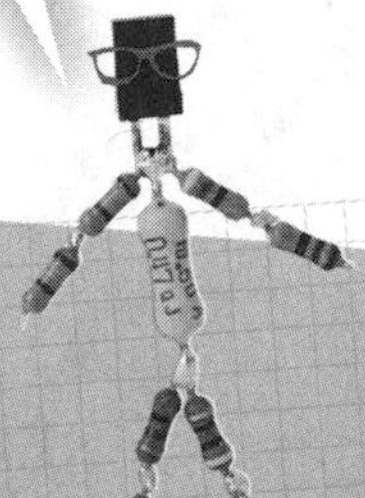

MAKE IT

1. GATHER KEYS

Pop some keys off the laptop with a screwdriver. This shouldn't require much force, so use a light touch.

2. REMOVE THE BUTTONS

Snip off the shirt buttons that you would like to replace. (Don't forget the cuffs!)

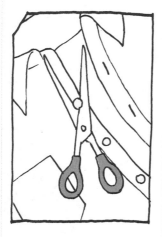

3. WIDEN

Widen the buttonholes, if necessary, by snipping the corners of the slits a little wider in each direction. Then clean up the widened buttonholes (and prevent them from unraveling) by finishing them with the aptly named buttonhole stitch (see right for a tutorial).

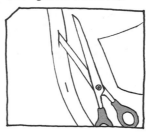

4. TAPE

Take a piece of tape and make a 6" loop so that the sticky side faces out. Press this loop down onto a piece of scrap wood. Stick the keys on top of the loop to keep them in place.

5. DRILL

Mark and then drill two vertically centered holes through each key with a ¹⁄₁₆" drill bit, about ⅛" apart.

Buttonhole, buttonhole, who's got the buttonhole?

To sew a buttonhole stitch, draw the needle up through the back, cross it over the thread pulled through from the previous stitch, and once again cross it over the thread as it comes out the front (which forms a loop). You pull tight and continue, repeating this pattern until you have gone completely around the buttonhole.

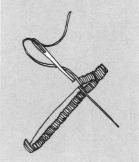

continued >>

6. SEW

Sew the new buttons onto the shirt in place of the old ones (keeping the spacing the same). Draw the needle through the fabric from inside to outside, then through one of the button holes (through the back); send the needle back down through the second hole and through the shirt fabric in nearly the same location that you came up. Repeat, reinforcing the stitches. To finish, wrap the needle and thread around the bundle of threads between the button and the shirt a couple of times. Then thread it through the fabric to the inside of the shirt and tie off the thread.

7. GEEK OUT

Try on your new shirt and geek out!

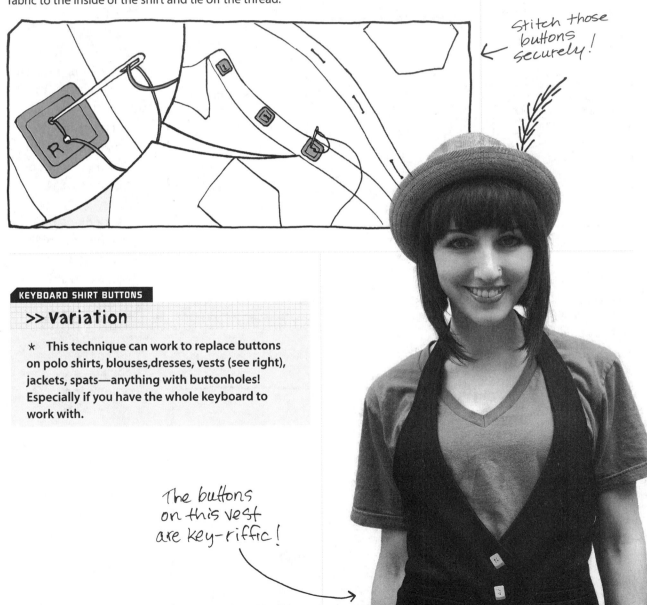

Stitch those buttons securely!

KEYBOARD SHIRT BUTTONS

>> Variation

★ This technique can work to replace buttons on polo shirts, blouses, dresses, vests (see right), jackets, spats—anything with buttonholes! Especially if you have the whole keyboard to work with.

The buttons on this vest are key-riffic!

#22 Hearmuffs

LEVEL 2: Intermediate | **TECH TRASH:** Headphones

The Vikings had many options in the afterlife, but the two big ones were Valhalla and Niflhel. In Valhalla, those who died heroically in battle would arrive to drink, fight, and be merry for eternity. If you ended up in Niflhel, on the other hand, you would spend an eternity in frozen mediocrity. (Niflhel, FYI, was a hamlet of Niflheim, the deepest realm of the Viking universe, a gray, limitless expanse of darkness, ice, and mist.)

The point is, be prepared for anything. Should it turn out that when I die a peaceful death on life's battlefield I end up going to Niflhel, I want to be ready. And I'm packing earmuffs—I mean, imagine spending an eternity in the frozen mist with cold ears! Convert a pair of dead headphones into comfy earmuffs that I can bring on my journey. *Note:* For those of you who aren't preparing for the possibility of a pagan afterlife, these Hearmuffs will also keep your ears warm in any local inclement weather!

MATERIALS

- Broken headphones
- Scissors
- Blank CD
- ½ yard warm, furry fabric
- Permanent marker
- Sewing needle and thread
- Polyester fiber stuffing

MAKE IT

1. CUT THE CORD

Well, it was fun while it lasted, but it's now time to cut the lifeline from your headphones. Trim the wires as close to the headphone as possible.

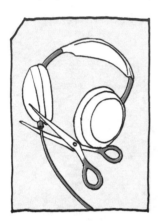

2. TRACE

Use your CD as a template and trace its outline four times onto the backside of your fabric. Cut out the four circles.

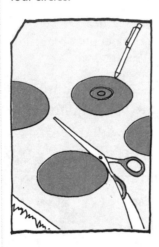

3. SEW

Sandwich the circles together in pairs, with the furry sides of the fabric facing each other. Carefully sew around the edge of each circle pair, leaving an opening at one side that is just wide enough for the headphone's earpiece.

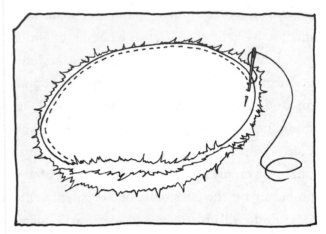

4. FLIP OUT

Turn the fabric right side out through the hole in the side (so the furry side is facing out and the stitching is inside).

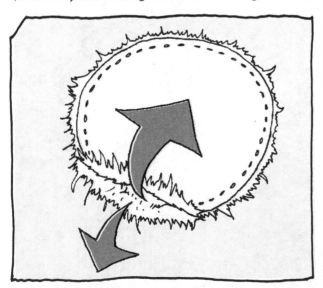

5. FILL

Fill each of the two furry little pockets with stuffing to medium density (the pockets should be able to keep their shape without being too firm). Then insert one of the headset earpieces into one of the pockets and hand-stitch around the opening to close the hole around the headset. Repeat on the opposite side.

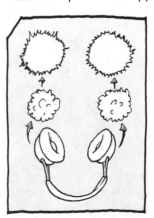

Rocking
Remote cuff
p. 95

Sustainable
Vase p. 77

cable cord coaster
p. 50

Obligatory
Decorative
candleholder
p. 163

DIY dinner
and a movie!
Deck out your
table and rig
a projector
for at-home
cinema.

DIY Digital
Projector
p. 203

Magnetic Memo Board (& keyboard magnets) p. **70**

Mouse Pencil Sharpener p. **146**

CD-Drive Desk Organizer p. **72**

Keyboard Pinwheel p. **127**

Drive Bookends
p. 75

Gear clock
p. 56

Music Monster p. 176

All-in-One
Shadow Box
p. 148

My First
Squiggle Bot
p. 131

Make
mice and
monsters
and robots,
oh my!

capacitor
Earrings p. 92

Polarized Wallet
p. 151

Keyboard
Buttons
p. 104

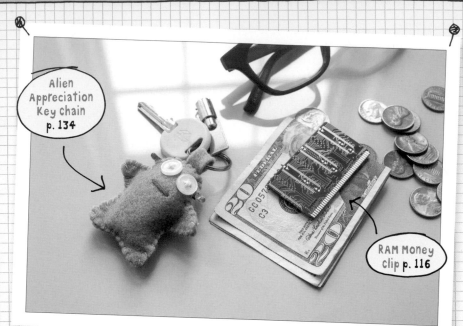

Techno-to-go-go: pocket-sized hacking.

Musical Hold
Box p. 224

Scanner
Compost Bin
p. 63

Postindustrial
Night-Light
p. 228

The ultimate
green machine:
Recycle your
scanners <u>and</u>
banan-ers!

Mr. Resistor Man
p. 129

#23 cable corsage

LEVEL 2: Intermediate | **TECH TRASH:** Ribbon Cable

At first glance this may seem like a very girly project, but the cable corsage was actually included in the book specifically for those geeky teenage guys out there. Sooner or later you're going to be asking an equally geeky young woman to a school dance. And you have to show up with a corsage. You may think the angel of your affections won't be into such a thing—after all, she made her own dress out of duct tape and shopping bags. This line of reasoning is dangerous.

Sure, it's easy to go to the local florist, but it isn't going to do much to win her heart. On the other hand, you could rip your computer apart and make her something that matches her personal style. You'll be amazed how far a small gesture and some elbow grease can take you in the quest for love. This lesson might not make much sense right now, but keep it stored in the back of your mind, because someday it will come in handy. I'm glad we had this talk.

MATERIALS

- Ribbon cables from a dead computer
- Scissors
- 1' stranded wire
- Diagonal cutters
- 1½" pin back
- Hot-glue gun

Mr. Resistor Man Says:
It is estimated that in 2005, one out of every eight couples who tied the knot met online.

MAKE IT

1. REMOVE THE CABLES

Remove the ribbon cables connecting the various drives to the motherboard from inside the dead computer.

2. REMOVE THE CONNECTORS

Remove the plastic connector boxes from the ends of the ribbon cables by breaking off the tabs on each side and pulling them apart. Set aside at least one connector to use in Step 10.

3. CUT THE CABLE

Cut four 2" to 3" strips of cable from one of the larger cables and fold them in half lengthwise.

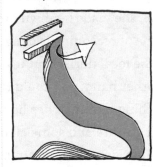

4. CUT

Take one folded strip and round the corners as shown. Repeat on the remaining three strips.

5. TIE

Fold each of the four cable strips in half so that the rounded edges meet. Arrange them on your work surface so the four petals are laid out like a clover, as shown. Pass a small piece of wire underneath two opposite "petals" and up through the folds, binding them together.

6. TWIST

Twist the ends of the wire in a half turn (as if you were wrapping a ribbon around a gift) and pass them down through the folds of the remaining two petals.

7. FANCY

Delicately flip the "flower" over, pull the wire tight, and twist the ends on the bottom. Knot and tie it to hold it all together. Make the center a tad more fancy by tying a large knot, adding another color wire, and/or spiraling the ends of the wire you used to tie the center knot.

8. DOUBLE IT

Make a second, smaller flower (cutting the ribbon cable ½" to 1" shorter than your first one), following Steps 3 through 7.

9. CUT LEAVES

Draw and cut two "leaves" (each about 3" long) out of the ribbon cable. (Use a slightly darker-colored cable if available.) Move around the flowers and leaves until you have an arrangement you are pleased with.

10. GLUE

Apply a generous line of hot glue along one flat side of the connector you set aside in Step 2 and press the flower arrangement on top of it. (*Note*: If the connector is not completely flat, cut off any tabs first so that it is.)

11. PIN

Hot-glue the pin back to the back side of the connector. And you're ready! This is about as far as my directions can take you. Pinning it to her duct-tape dress is your problem.

Have no fear of wilting petals or seasonal allergies with these flowers!

#24 8-Bit BELT Buckle

LEVEL 2: Intermediate | **TECH TRASH**: Nintendo controller

Some things are built to last. NES controllers are one such thing. They are simple, elegant, and virtually indestructible. Unfortunately, NES consoles . . . not so much. If you don't believe me, here's an experiment: Pull your old NES console out of the garage, dust it off, hook it up to your TV, and try playing.

Face it, no matter how many times you blow into that darn cartridge, you will never get *Metroid* to load right. Perhaps it is time to part with your Nintendo and show appreciation for it in other ways. Which brings us back to your perfectly functional controllers. Commemorate your old Nintendo Entertainment System by turning the controller into a blinged-out belt buckle. Nothing says "I love my old-school Nintendo" quite as loudly as a shiny silver and gold controller belt buckle. If you have two controllers, you can make a pair of his-and-hers buckles!

MATERIALS

- Nintendo controller
- Phillips-head screwdriver
- Pliers
- Marker
- Power drill with a 1/4" drill bit
- Diagonal cutters
- Wire coat hanger
- Blue painter's tape
- Respirator
- Craft knife
- Metallic silver spray paint
- Metallic gold spray paint
- Hot-glue gun
- One 32 by 1/2 machine screws (with nuts)
- Belt (with snaps for a belt buckle)

Mr. Resistor Man Says:
In its heyday, the NES gaming console sold more than 61 million units and held around 90 percent of the 8-bit gaming market share.

MAKE IT

1. DISASSEMBLE

Open the controller using your screwdriver. Remove the electronics and the buttons. Set aside the screws for later reassembly.

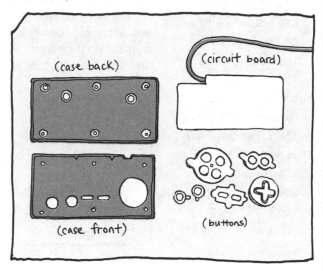

2. MARK IT UP

On the back of the controller casing, make two marks 1¼" in from the right side of the case, aligned with the rows of screw sockets. Make a third, centered mark about 1" in from the left side of the case. Then use your power drill to make holes through the casing at each of the three marks.

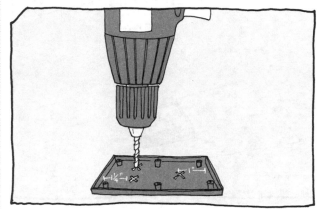

3. HANGER LOOSE

Use your diagonal cutters to trim a 6" piece of wire from a solid metal hanger. Then bend the wire so that it forms a rectangular U shape.

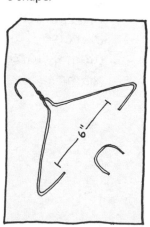

4. FIT

Insert the "legs" of the U through the controller at the parallel holes you drilled in Step 2. Trim and bend the ends of the wire on the underside to form a roughly 1¼" square.

5. TAPE IT

Cover the face of the front panel with tape, then use your craft knife to carefully cut around the buttons. Peel away the tape that covers the buttons to expose the gray trim.

continued >>

6. PAINT

Put on your respirator, take your controller outside, lay it on newspaper, and spray-paint several light coats of metallic silver. *Note*: It's better to put on many light coats than one heavy coat.

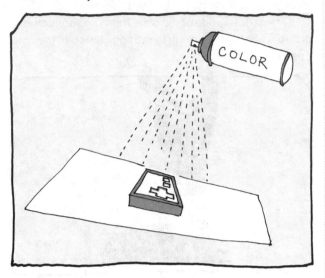

Extra credit!

If you are feeling really inspired, you can also use golden paint on the tops of the screws used to close the controller. Just remember to cover the threading with tape so that it stays paint-free and you can close the controller later.

7. PEEL

Once dry, peel the tape off the front of the controller. Apply more tape to the front panel so that you cover the painted silver area. This means you will need to cover the trim around the buttons and once more cut it away with a craft knife.

8. PAINT AGAIN

Take the controller outside and spray-paint the front with metallic gold. Paint the buttons and arrow pad gold as well. Once the gold paint is dry, carefully peel off the tape.

9. GLUE

Insert the buttons and keypad back into the controller. With your hot-glue gun, stick these buttons in place at the back of the controller's front piece.

Hot-gluing is much better exercise than playing video games.

10. HOOK

Take your machine screw and thread it through the center hole in the controller back, inside to outside, so the end of the screw sticks out the back. If the screw appears to be loose, fasten it on with a nut. If not, it's not loose, so don't worry about it!

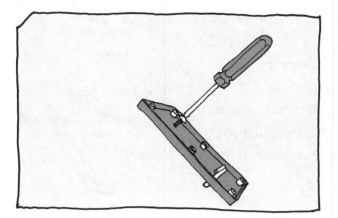

11. REASSEMBLE

Use your screwdriver to reassemble the controller by fastening the case shut with the screws you removed in Step 1.

12. ATTACH THE BELT

Pass the belt through the wire hanger loop and snap the belt in place. It is now ready to wear.

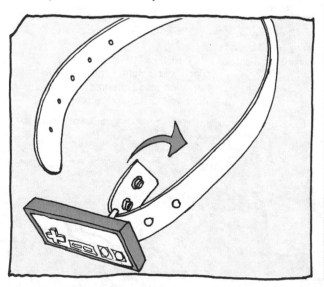

#25 RAM Money Clip

LEVEL 3: Expert | **TECH TRASH**: Dead Computer

I know I am supposed to diplomatically claim that every project in this book is my favorite, but can you keep a secret? This just may be at the top of my list. Since having made my own RAM Money Clip, it has been with me every single day, holding my money, receipts, and other loose papers. And if merely keeping my pockets tidy weren't reason enough for it to become my new BFF, it's stylish, fun, and quite handsome. I love that people always ask me where I got it, and that no one ever believes that I made it myself or that it is, in fact, so easy to make.

MATERIALS

- Dead computer
- Heat gun
- Pliers or screwdriver
- Ruler
- Blank metal money clip*
  *Blank money clips can be purchased from jewelry supply stores, craft stores, and other online suppliers.
- Large plastic container
- Nitrile gloves (available at any pharmacy)
- Scissors
- Sandpaper
- Rag or paper towel
- Soldering setup
- Exhaust fan

It's all about the "Jacksons".

Mr. Resistor Man Says:
Since gold is highly conductive and corrosion-resistant, some circuit boards have gold-plated connectors.

MAKE IT

1. DIGITAL LOBOTOMY

Open the side of the computer and locate a RAM card with memory chips on just one side (as opposed to two). It is a little flat circuit board covered in microchips that is plugged into your motherboard. Remove the RAM card by unplugging it.

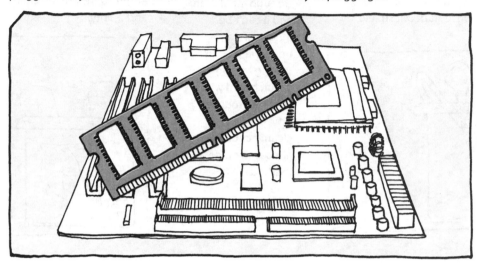

SAFETY FIRST

>> Choosing a RAM card with memory chips on one side is not only for aesthetic, it's for safety, too. If it has chips on both sides, both sides have solder on them, as well. Solder often contains lead, which shouldn't be rubbing around on the stuff in your pocket.

2. DESOLDER

Use your heat gun in a well-ventilated area (preferably with a concrete floor) to heat up and remove components from the board. Place your circuit board on the floor and blast it for a few seconds with the heat gun. Pieces are ready to be removed when the solder starts to "silver" and they start falling off on their own. If they aren't falling off but are loosened, use a pair of pliers or long screwdriver to pick or push the components off while the board is still hot. Wait about five minutes for it to cool before you handle it.

SAFETY FIRST

>> If you ever do this project with a circuit board other than a RAM card, remove all batteries and electrolytic capacitors before applying heat or they will explode and possibly harm you.

While this project is very easy, safety is key.

continued >>

3. CUT

Measure the size of the money clip (it should be about 2" by 1"). Then fill a large container with water and put on the nitrile gloves. Submerge the RAM board and, with a pair of scissors, cut it (underwater, to prevent dust) to the dimensions of your money clip. It may take some force to get the cut started, but once started, it should cut like paper.

4. SAND

Keep your gloves on and submerge the board once more. Use a small piece of sandpaper to round the edges so they're not sharp and pointy.

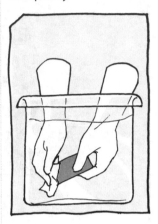

5. WASH AND SOLDER

Wash the board under running water so that any leftover dirt is cleaned off. Dry it well with a rag or paper towel. Then apply a few drops of solder evenly around the edges of the top of the money clip. Also, apply a little solder to the underside of the circuit board.

6. HEAT

Arrange the circuit board facedown on a concrete surface (or surface that is heat-resistant). Align the money clip so the top is centered on the circuit board. Keeping your hands clear, blast your money clip with a heat gun until the solder liquefies and the clip visually appears to settle atop the circuit board.

7. WAIT . . . FOR THE MONEY

Let the money clip cool for at least ten minutes. While you're doing that, wash the solder off your hands. Once your money clip has cooled off, fill it with all of your precious cash.

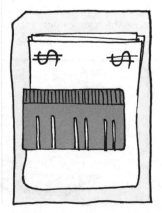

#26 The DON'T DIE-POD

LEVEL 1: Novice | **TECH TRASH:** Broken iPod

During its life, your iPod went everywhere with you, but now in death, it remains buried in your drawer, along with the hope that perhaps someday you will discover a way to bring it back to life. Though these must be trying times for you, dear reader, I have some words of consolation: With a little bit of imagination and a lot of "surgery," we can resurrect your iPod to give it a new, purposeful life. From now on, your iPod will do much more than keep you entertained. Your iPod will be your guardian angel, there to keep you alive (or at least with a healthy supply of bandages) no matter what life throws your way. Even tiger attacks.

MATERIALS

- Broken iPod
- Razor blade
- Mini screwdriver set
- Hot-glue gun
- Scissors
- Elastic shock cord
- Diagonal cutters
- Red construction paper or electrical tape
- Clear packing tape
- Torx wrench (optional)
- First aid supplies

Mr. Resistor Man Says:
In less than ten years, Apple manufactured 17 different versions of the iPod MP3 player.

MAKE IT

1. OPEN THE CASE

Opening the case of an iPod is the hardest part. Use a razor blade to create a small gap between the top and bottom halves. While carefully holding the gap open with the razor blade, use a flathead mini screwdriver and gently pry the two halves apart.

2. GUT IT

This is the most fun (and gut-wrenching) part. Remove the innards of your iPod so you are left with an empty case. (You may need to remove a few screws and unplug some cables, but most everything should just pull out.)

3. NOW YOU SEE IT

Use your screwdriver to disassemble the scroll-wheel mechanism. With your razor blade, carefully pry apart the circuit board from the scroll wheel. (You need to remove the circuit board so there's no lead in our first aid kit.)

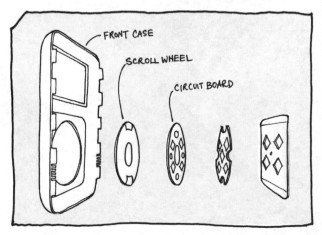

FRONT CASE

SCROLL WHEEL

CIRCUIT BOARD

4. NOW YOU DON'T

Once the circuitry is removed, reassemble the scroll wheel and hot-glue it in place.

5. PLASTIC SURGERY

Carefully remove the plastic tabs from the front of the case so that only the two tabs on the bottom of the case and the two on the bottom right side remain.

6. KNOT IT

Cut a 12" piece from the shock cord. Pass one end of the cord (from outside to inside) through the hole for the headphone jack and the other end through the adjacent hole in the case. Tie the cord ends on the inside of the case in an overhand knot, leaving about 2" at the ends. Trim the ends once the knot is tightened. (This cord will become a wrap to hold the case shut.)

7. LABEL

Cut a small red cross from the construction paper, using the dimensions of the iPod screen as a guide. Tape it to the *inside* window of the case using the clear packing tape. Trim the excess tape. *Note:* Red electrical tape, fabric, or red laminate can also be used in place of construction paper and packing tape.

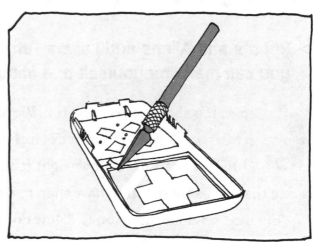

8. FILL 'ER UP!

Fill up your Don't Die-Pod with all of your favorite first aid/survival tools.

Load it up!

Here are some suggestions for filling your Don't Die-Pod case:

* Band-Aids
* Alcohol swabs
* Disinfectant wipes
* Gauze pads
* Q-tips
* Tweezers
* Sports tape (can be wrapped around the inside of the case)
* Ibuprofen
* Tampons
* Condoms
* Tourniquet (for tiger attacks)
* Emergency cash (to get yourself to the hospital after a tiger attack)
* Prepaid phone card

CHAPTER 4

Fun-Fun-Fun-Fun-Fun-Fun!

> **Robots and Aliens and Lasers—oh, my. Here are 6 clever projects you can make for yourself or a friend—just for the fun of it!**

L ife is too short not to have fun. We all love fun! Yet, what is it? Can it be easily explained? No, because trying to define "fun" is not fun. It is downright tedious. How about we just agree that fun is more fun to have than to define?

The next obvious question is, "How do we have fun?" "We know it when we're having it" is a popular answer. Then again, how do we know we're having it? Perhaps it's because we are having a good time and not just any good time, but one that is whimsical and spontaneous and pleasant. Perhaps fun is a combination of circumstances that result in an enjoyable, whimsical, and possibly spontaneous situation.

Mr. Resistor Man Says:
On April 30, 1952, Hasbro's Mr. Potato Head became the first toy to be advertised on American television.

My First Squiggle Bot, p. 131

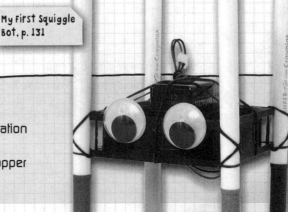

PROJECTS

And while it would be foolish of me to guarantee enjoyable, whimsical, and slightly spontaneous times (after all, fun is based on perception and is not a universal attribute), I feel comfortable enough to assert that making these projects could possibly, maybe—likely—be fun. You will learn new things, work with your hands, and discover the unexpected. Assuming that you enjoy these experiences, these projects may very well result in an enjoyable and uplifting afternoon.

All of that said, this chapter alone isn't about having a fun time making projects—this *book* is about having a fun time making projects—this chapter in particular is about making projects that are factories of fun. You will make things that will be so exceedingly fun that your head may just explode from the muscle tension created by your immense prolonged smile. Let your fun extend well beyond an afternoon of tinkering into many afternoons of play by making things that are enjoyable to play with for a long time!

Alien Appreciation Key Chain, p. 134

After all, when we look at electronics, especially broken electronics, we are probably more likely to see dull, tired garbage and the fun times we had along the Oregon Trail rather than the fun we could still be having. This doesn't have to be the case. We can strike out from camp and forge another path through uncharted wilderness. It merely requires a little imagination to get started on our journey. Next time you look at "garbage," perhaps you don't see a chore laid out in front of you, but good times yet to be had.

PDA Doodler, p. 124

Keyboard Pinwheel, p. 127

Mr. Resistor Family, p. 129

#27 PDA Doodler

LEVEL 2: Intermediate | **TECH TRASH:** Broken PDA

Imagine this: It's Tuesday, you're at another dull company meeting, diligently "penciling" notes into your personal planner. While your boss thinks you are gearing up for a regional sales push, you are in fact drawing a fantastic likeness of the district manager, who is sitting across the table from you, with a kickin' Mohawk.

You then survey your handiwork and chuckle to yourself. The room gets silent. Everyone turns and looks at you. Glancing up, you explain that you laughed out loud just thinking about how the new sales initiative being outlined will exponentially increase shareholder value. Everyone looks pleased by your answer. The meeting resumes, and you go back to your notes. Thanks to a little bit of creativity and initiative, you're now going places in this company!

MATERIALS

- Broken PDA
- Mini screwdriver set
- Hot-glue gun
- Razor blade or utility knife
- Pencil
- Black mat board
- Straightedge
- Craft knife
- Double-sided tape
- Black and white drawing on photo paper (Tip: Draw an image by hand, take a picture, and have it professionally developed on photo paper.)
- Black marker
- Iron filings
- Epoxy
- Small cylindrical disc magnet
- Black mat board

MAKE IT

1. GUT IT

Remove the screws from the back of the PDA. Open the case and remove the electronic components. Set the screws aside for reassembly.

2. GLUE

Using your hot-glue gun, glue the buttons in place on the outer case (sealing any holes). Set aside.

3. DISASSEMBLE

Cautiously separate the PDA's front glass panel of the PDA from the LCD assembly with a razor blade or utility knife.

4. START DRAWING

Use the window from the front of the case as a guide and trace along its inner edge onto the white side of your mat board. These will be the inner cut lines for your frame.

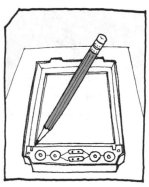

5. FINISH DRAWING

Lay the glass panel over the marked inner cut lines so that you have roughly an equal margin on all sides. Use the edge of the glass to trace the outer cut lines for your frame.

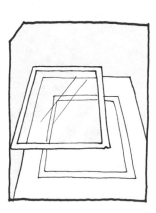

6. CUT IT OUT

Using a straightedge to follow the marks you just traced, cut out the frame with a craft knife. Check that the frame's outer edge is the same size as the glass and that the inner edge is aligned perfectly with the PDA's front window when placed inside the case.

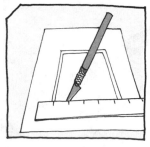

Some black mat board is only black on one side. If this is the case, make certain the black side is facing up.

HELPFUL HINT

Color in the white edges on the inside of the frame with a permanent marker to make your end result look cleaner and more finished.

7. MOUNT

Apply double-sided tape around the edges on the back of your mat board frame. It's okay if the tape sticks off the sides a little, but keep it away from inside the frame. Center the black and white drawing in the frame and then firmly press it down.

continued >>

8. TRIM

With your craft knife, cut away any photo paper sticking out from the outer edge of the frame.

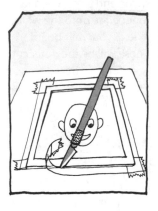

9. FILL

Carefully fill the frame with a pinch of iron filings. Place the glass over the frame and gently test it out by picking up and moving around filings with a magnet. Add another pinch of iron filings if needed (and subsequent pinches, if needed, keeping in mind it's easier to add filings than subtract).

10. GLASS

Apply double-sided tape to the border of the frame and press the glass over it. Again, be careful not to get tape in view of the window. Trim away any excess tape that extends from the frame.

11. GLUE

Glue the entire glass assembly into the front of the case with the hot-glue gun and reassemble your PDA with the screws you set aside in Step 1.

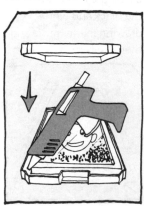

12. CUT AND GLUE

With a craft knife or pair of scissors, cut the tip off the little touch pen that comes with the PDA. Then, using high-strength epoxy, attach the magnet to the tip of the pen.

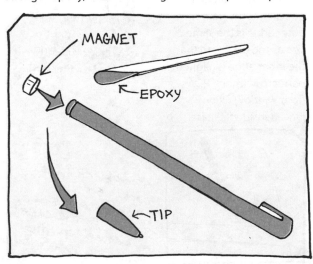

MAGNET

EPOXY

TIP

13. LIGHTEN UP

After the epoxy is dry (follow the instructions on the packaging), slip your new doodler in that old leather PDA. Forget the two o'clock meeting and start doodling. Just don't get caught!

#28 KEYBOARD Pinwheel

LEVEL 1: Novice | **TECH TRASH:** Keyboard

Wind is one of the most abundant resources on planet Earth. And if you judge something's worth based on its abundance (the way my grandparents judge the food at restaurants), then wind is *spectacular.* But while engineers are figuring out how to put the wind to work to solve the world's energy crisis, let's use our efforts to give wind the day off for some fun. These pinwheels are for the wind to play with. After all, maybe you'll have a little fun in the process, too.

MATERIALS

- Keyboard
- Phillips-head screwdriver
- Marker
- Ruler
- Craft knife
- Straight pin
- Hot-glue gun
- Drinking straw
- Diagonal cutters
- Piece of tape

MAKE IT

1. OPEN THE KEYBOARD

Remove any screws from the bottom of the keyboard and pry it open.

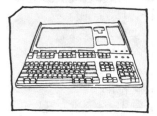

Pry it open, gently but firmly.

2. LOCATE THE PLASTIC

Find the flexible, transparent plastic circuit board inside the keyboard. Remove it, and mark a 4⅛" square. Cut a 4" square just inside the marks.

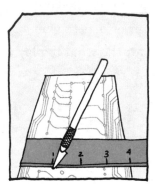

3. MARK AND CUT DIAGONALLY

Bisect the square diagonally with your ruler. Mark and cut just 2" in from each corner toward the center.

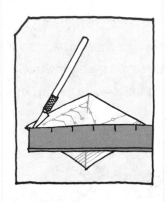

4. FOLD

Gently bend every other corner toward the center point without creasing the plastic. You should see the pinwheel take shape.

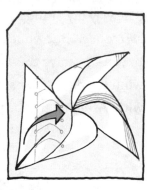

5. GO FOR THE PIN

Insert a pin through each of the four layered corners that you just curved inward and then through the center of the pinwheel.

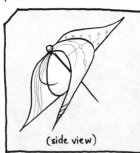

(side view)

6. GLUE

Apply a drop of hot glue to the pin in the front and back of the pinwheel to keep it from falling out.

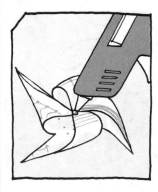

7. ATTACH THE STRAW

Insert the pin point through two layers at the end of the straw. Clip off the point of the pin with your diagonal cutters. Fold a small (about ¾") square of scotch tape over the clipped end of the pin.

8. THAR SHE BLOWS!

Go outside and let the wind do its thing. Hold onto your hat!

Note: If the wind is having an off day, create your own. Just put your lips together and blow.

#29 Mr. RESISTOR MAN

LEVEL 1: Novice | **TECH TRASH:** Circuit boards

I f you have made a couple of projects already, like the Gear Clock (page 56) and the Walkman Soap Dish (page 86), it's likely you've already amassed a fair-size collection of unused circuit boards. Some of the components on them probably still work. However, if you're not a budding engineer and they all look like little doodads and ziggy-zags, use these parts to make cute little creatures (and, perhaps, familiarize yourself with the components in the process). One such cute little creature is this contemporary classic, Mr. Resistor Man.

MATERIALS

- ▸ Circuit boards
- ▸ Diagonal cutters
- ▸ Lead-free solder
- ▸ Soldering iron

Mr. Resistor Man Says:
During recycling, circuit boards are ground into a powder and separated into metals, precious metals, and fiberglass.

Ms. Resistor Woman Says:
Resistors are used in circuits for providing resistance to the flow of electricity and converting electric energy to heat.

Fido, the Resistor Dachsund Says: Woof.

MAKE IT

1. COLLECT COMPONENTS

Resistors are little ceramic and plastic tubes, often striped, with wires sticking out each end. Once you've found some, use diagonal cutters to snip them and other components (LEDs and small capacitors) off the circuit boards.

2. PICK A BODY

After gathering 12 to 20 small components, choose a large resistor (A) to use for the torso.

3. HEAD

Select a head (try an LED or small capacitor) and line it (B) up with one end of the torso resistor.

4. SHOULDERS, KNEES

Select four matching resistors for the arms (C) and four for the legs (D).

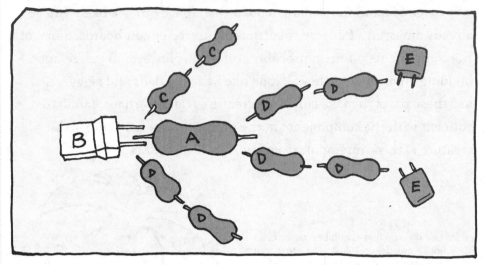

5. AND TOES

Find two matching flat capacitors for feet (E). Ideal feet are large and flat, because this will later help Mr. Resistor Man stand upright.

6. SOLDER

One joint at a time, solder each of Mr. Resistor Man's parts together, from head to toe.

7. GET UP, STAND UP

Gently bend the feet (capacitors) forward at a 90-degree angle so that Mr. Resistor Man will be able to stand on them. Position him so that he can stand on his own two feet.

Be patient, it may take some practice for Mr. Resistor Man to get his footing.

MR. RESISTOR MAN

>> Variation

★ Make a Ms. Resistor Woman, a pet dog, a baby, an elf, a rabbit, an antlered deer, a unicorn, a writhing octopus—and whatever else you can imagine. Make a resistor family, an entourage, an army to take over your desk, the cast of *Star Wars* or *Romeo and Juliet*!

#30 MY First SQ

LEVEL 2: Intermediate | **TECH TRASH:** Computer Fan

L ife is about baby steps. As much as you'd like to go
a robotic Rembrandt, you're going to have to take i
Robots don't just mature into masters overnight. Like w
is important to let robots grow up at their own pace. All r
learn to squiggle before they can draw. So, let your newbor
explore the page freely. It may just surprise you!

MAKE IT

1. CHOP OFF THE BLADES

Use your diagonal cutters t
the fan blades in a wa
next to each other
clip off six that
that are si

arkers
▸ 9V battery clip
▸ Hot-glue gun
▸ Googly eyes
▸ Large sheet of
  paper

Mr. Resistor Man Says:
The word "robot" was
invented by a Czech
writer in 1921 for a
stage play in which a
factory manufactured
artificial people.

... clip off about two thirds of ... that the remaining blades are all ... For instance, if there are nine blades, ... are clustered together, leaving three blades ... by side.

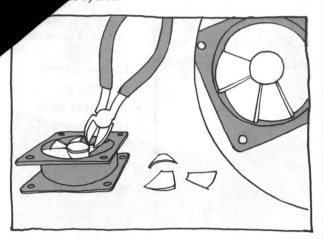

2. MOUNT THE BATTERY

Center the battery on the stationary part of the fan. Wrap two to four rubber bands in both directions around the whole contraption to hold it steadily in place.

3. ADD RUBBER BANDS

Pinch one rubber band and thread it through the mounting holes at the corner of the fan, forming two loops (one on either side of the holes). Repeat with three rubber bands on the remaining corners.

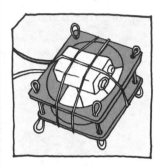

4. ATTACH THE MARKERS

Wrap the two loops of each rubber band around a capped marker several times, securing a different-color marker in each corner. Make sure the marker tips (the capped ends) are facing down (on the opposite side of the fan from the battery). When you are done, the markers should look like four table legs.

HELPFUL HINT

To better hold each marker in place, twist the end of the rubber band farthest from the cap and pull it around over the cap.

5. PREPARE AND KNOT THE WIRES

Line up the ends of the wires from the fan and the ends of the wires from the 9V battery clip so the battery clip wires extend an inch farther. Then tie the wire ends in an overhand knot. After the knot is made, the wires from the battery clip should still extend farther than those from the fan.

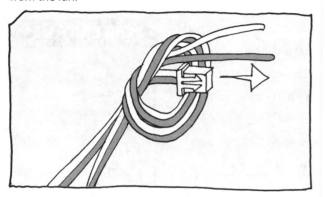

6. WIRE IT UP

Plug the red wire from the battery clip into the fan's wire socket so that it lines up with the red wire. Repeat with the black wire.

7. GOOGLY EYES

A Squiggle Bot would not be complete without googly eyes. Apply a small amount of hot glue on the back of the googly eyes and press them next to each other on one side of the Squiggle Bot frame.

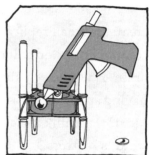

8. SQUIGGLE AWAY!

To start squiggling and doodling, place a large sheet of paper on the floor, remove all of the marker caps, and plug in the battery.

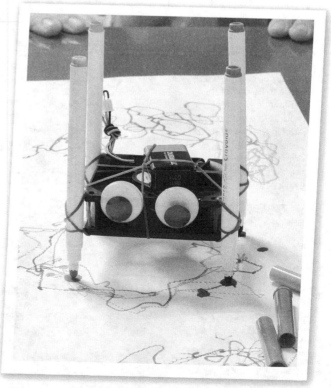

#31 Alien APPRECIATION Key Chain

LEVEL 3: Expert | **TECH TRASH**: Old Computer

This key chain should prove useful the next time you are abducted by aliens. The little buggers will surely appreciate that you have created a small replica of them using primitive photon-emitting active components, and will perhaps stop embedding transmitters in your brain. Having been abducted two-and-a-half times myself (and once having spent an afternoon in the UFO Museum in Roswell, New Mexico), I think it is fair to say that you can trust me on these very serious matters.

Mr. Resistor Man Says:
Hundreds of thousands of copies of *E.T. the Extra-Terrestrial,* considered by some to be one of the worst video games ever made, were buried by Atari in the New Mexico desert (appropriately, not far from Roswell) during the Video Game Crash of 1983.

MATERIALS
- Old computer
- Phillips-head screwdriver
- Scrap paper
- Marker
- Felt
- Scissors
- Sewing needle
- Thread
- Two four-hole buttons
- Two colors of embroidery thread
- Wire cutters
- Resistor (100 ohm)
- Soldering setup
- Aluminum tape
- Gaffer's tape (or duct tape)
- Fiberfill stuffing
- Key ring

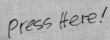
Press Here!

MAKE IT

1. GET YOUR LEDS

Open your computer case by removing the side panel. Next, remove the front panel and detach the LEDs from the case. Finally, unplug the LEDs from the motherboard.

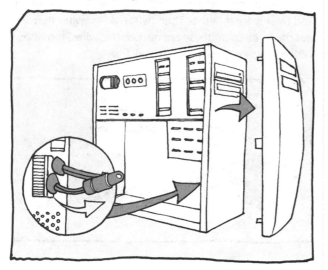

2. GET YOUR BATTERY

The motherboard should have a 3V coin cell battery. Remove this battery to use as the power source for the LEDs.

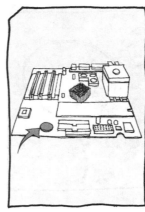

3. MAKE A PATTERN

Make a small pattern no more than a few inches tall that has a head large enough to pass your battery through, a body large enough to hold your battery, two tiny arms, and two little stumps for legs, as shown.

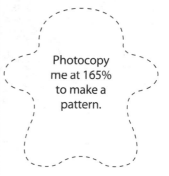

Photocopy me at 165% to make a pattern.

4. TRANSFER

Fold a piece of felt in half. Cut out your pattern from the sheet of paper and, with a marker, trace it onto the felt. Cut along the traced lines, through both layers, to make two equal felt cutouts.

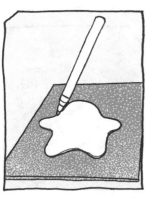

5. SEW!

Slide the piece of felt with the marker on it to the back so that once you start sewing, the marker lines will be on the inside. Thread your needle and tie the ends in a knot. Start sewing at the neck by stitching from the inside of the fabric out until the knot catches. Continue sewing a whipstitch around the edge of the body, until you reach the other side of the neck. Knot the thread so it doesn't unravel.

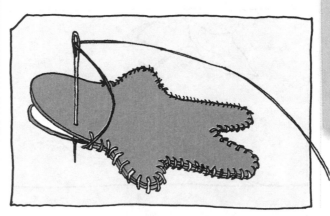

HELPFUL HINT

The whipstitch is an over-edge sewing technique. You stitch always pulling the needle through from the same side and advancing while you go. This will generate a diagonal-looking stitch on the outer seam.

continued >>

6. EYE CAN SEE

Position buttons on your key chain where the eyes should be. Using embroidery thread, sew these to the front piece of the fabric, stitching to form an "X" on the visible side of the buttons.

7. A MOUTH

Using a different color embroidery thread, sew a mouth onto your key chain.

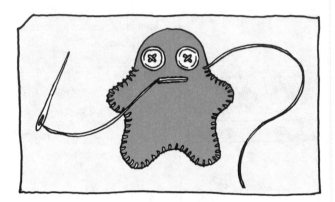

8. GET TECHNICAL

Trim the wires extending from your LED so that they are only about 2" to 3" long.

9. WIRE IT UP

Each LED should have a wire that is the same color. These wires are the ground wires. Strip off a little of the jacketing and twist them together. Then twist the two wires that are not the same color (these are the positive wires) together with one side of your 100-ohm resistor.

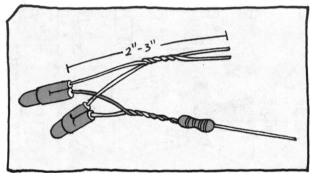

10. SOLDER

Solder together each of the two sets of twisted wires.

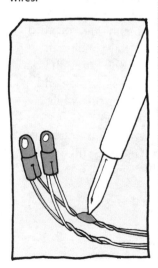

11. START WIRING

Bend the remaining resistor lead into a loop. Using aluminum tape, attach the wire loop to the positive side of the coin battery (the side with a plus on it). Be careful not to wrap the aluminum tape around the battery and have it touch the negative side, which would cause your battery to short (which is bad—it could get very hot and burn you, stop working, or explode).

12. CONTACT

Take a 2" piece of aluminum tape and fold one third of it onto itself, sticky sides together. Next, place the LED ground wires onto the folded part and fold the remaining third over so that the wire is stuck in place.

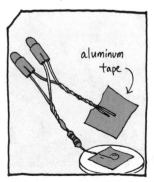

13. SPACING

Cut a small circle of felt slightly larger than the battery. Then cut an approximately ¼" hole in the center.

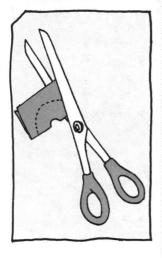

14. FINISH THE SWITCH

To finish the switch, sandwich the small piece of felt (with the hole) between the battery and the piece of aluminum tape connected to the ground. When you squeeze this together, the LEDs should light up, and they should turn off when released. Once you are sure this is happening, tape together the whole assembly with gaffer's or duct tape.

15. STUFF IT

Stuff the switch you made through the hole on the side of your little felt alien. Push in a few small pinches of fiberfill, too, to fill out the body and make it soft.

16. LOOPY

Cut a small strip of felt (about 1" by 1½") and fold it in half to make a key ring loop. Pinch it in place at the top of the alien head, between the front and back layers.

17. SEW SHUT

Position the LEDs about ½" on either side of the loop and sew the head shut in the same way you sewed together the body in Step 5, making sure to go through all the fabric layers to make them secure. Add an extra stitch *between* the wires of the LED on each side to help hold them in place.

18. FINISHING TOUCH

Insert the key ring through the head loop. Your new little friend will be outta this world.

To see your little alien in action, give it a little squeeze.

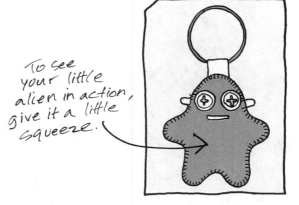

#32 Laser GAG Zapper

LEVEL 3: Expert | **TECH TRASH:** Broken NES Zapper

This laser gun won't vaporize or even cauterize, but that is the fun of it! Tell people that you turned your old NES zapper into a powerful laser gun, pick a target across the room (a prized possession perhaps), pull the trigger, and "BANG." The word "BANG" will be printed largely against the wall. It is essentially a modernized version of the classic gag prop gun that waved a little flag that said "BANG" when the trigger was pulled. This new version takes the joke that never gets old and carries it into a brave new millennium.

MATERIALS

- Laser pointer that uses 2 AAA batteries
- Pliers
- Phillips-head screwdriver
- Broken NES zapper
- Diagonal cutters
- Battery holder for 2 AAA batteries
- Mini screwdriver set
- Knife
- The word "BANG" printed on a slide or a transparency
- Marker
- Scissors
- Mat board
- Two-sided tape
- Hot-glue gun
- Fish-eye door peephole
- Soldering setup
- Wire
  - White sheet of paper
    - Epoxy

Mr. Resistor Man Says:
Did you know? The word laser is actually an acronym for "light amplification by stimulated emission of radiation."

MAKE IT

*Make Transparency

To create a transparency of the word "BANG!", first open up any text-editing software. Make the background of the page black and the text color white. Save this file to disk and bring it to any nearby copy center that prints files on transparency. Make sure the image on the transparency is no larger than ¾" across. Note: You can also scan the page in the back of the book and bring that file to the copy center.

1. FREE THE LASER

Take the laser pointer, pry or twist it open with pliers, and remove the laser diode from its casing. Take out all the electronics from the casing. *Note:* In some cases, this may require cautiously twisting it free with pliers.

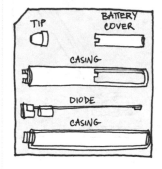

SAFETY FIRST

>> Lasers, even $5 laser pointers, can be dangerous. Never point them at yourself or others. Read and heed the warnings and handling procedures on the packaging.

2. OPEN THE GUN

Remove the screws on the NES zapper to open up the case. Set aside the screws for later reassembly.

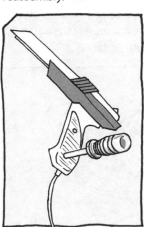

3. GUT IT

Cut the circuit board out of the gun, but leave the two wires connected to the tabs on the trigger assembly. Remove the lens in the front of the barrel and the weight in the handle. Set aside the lens somewhere safe.

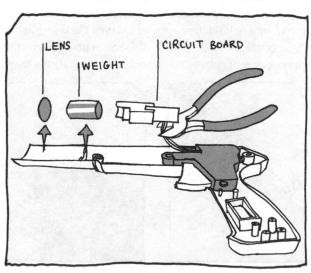

4. CLEAR IT OUT

With diagonal cutters, remove enough plastic from the inside of the handle so that the AAA battery holder will fit securely inside. *Note:* This may require removing a screw socket. Just don't remove more or the case may not close correctly.

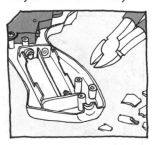

continued >>

5. SPY THIS

First, unscrew the two halves of the peephole that are screwed together. The piece that fastens on the outside of a door should have a small focusing lens toward the end where you would place your eye. Take a knife or thin screwdriver and pry this lens out from the cylinder.

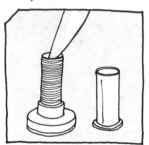

6. TRACE

Take the lens that was at the end of the barrel and center it over the word "BANG" on the transparency. Trace around the perimeter of the lens with a marker. Then use a pair of scissors to cut out the circle.

7. ANOTHER CIRCLE

Using the lens as a guide, once again trace and then cut out a circle, this time from a piece of mat board. Then cut the circle horizontally into thirds. Discard the center piece. Attach two-sided tape to each of the remaining pieces and stick them on the transparency above and below the word "BANG" as a frame.

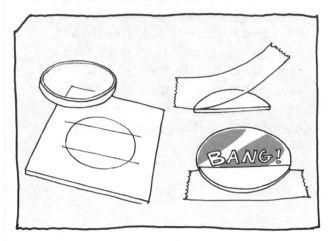

8. INSERT

Insert the "BANG" disc into the barrel of the zapper where the lens used to be and glue it in place, right side up and right-reading, with a small drop of hot glue. *Note:* If you were to look down the barrel of the zapper, the writing should appear backward.

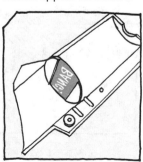

9. PEEPHOLE

Place the fish-eye door peephole into the gun so it sits between the hollow black weight in the barrel and the transparency. Use the weight to hold the tube extending back from the peephole in place.

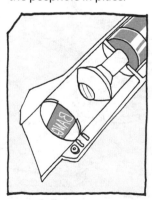

10. HACK

Connect the red wire from the AAA battery holder to the positive terminal on the laser assembly and the black wire to the negative. Locate the small surface mount switch on the circuit board and connect the solder pads at the base of the button with a piece of wire (stripped on both ends) until the laser turns on. Once you are sure which two solder pads turn on the laser, disconnect the battery holder, and solder the wire to both pads.

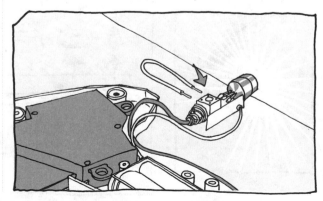

11. SWITCH

Solder the red wire from the AAA battery holder onto the terminal of the switch on the zapper that has the black wire connected. Now, cut away both wires that were originally connected to the switch.

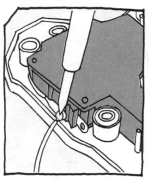

12. PREP

Return to the laser pointer assembly and solder a 3" red wire to the positive terminal and a 3" black wire to the negative terminal.

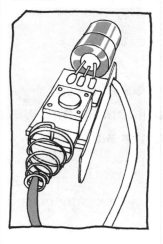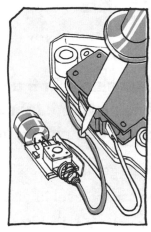

13. FINISH IT

Twist together the black wire from the laser assembly and the black wire from the AAA battery holder, then solder them together. Also solder the red wire from the laser assembly to the terminal on the zapper's switch that used to have a red wire connected (and currently has no wires connected).

HELPFUL HINT

Don't be afraid to readjust the laser while the glue is drying. The goal is to get the laser locked into the best position possible to shine through the lens.

14. TEST

Hold a white sheet of paper at the end of the barrel. Pull the trigger down halfway to shine the laser through the lens assembly. You should see the word "BANG!" shining onto the paper. Adjust the laser as needed for proper alignment and use a little hot glue to hold it in place.

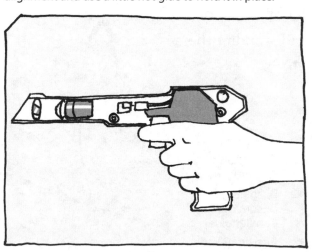

15. EPOXY

Once you are sure everything is aligned, use fast-setting epoxy to hold the laser, lens assembly, and "BANG" transparency in place. Let it dry overnight in a well-ventilated area.

16. REASSEMBLE

Reassemble the zapper and have a blast!

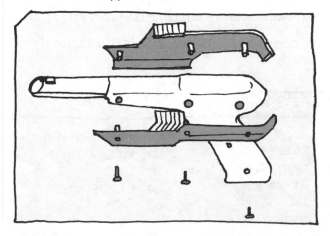

CHAPTER 5

Arts and CRAFTS

> **Indulge your creative side with 9 projects that turn computer scraps into sculpture (and notebooks and wallets and pencil sharpeners and picture frames and . . .).**

Sometimes I like to get in touch with my creative side. Yes, sometimes I truly revel in visually expressing myself through artistic creation. I'd like to think that at one point or another, everyone likes to express themselves through arts and crafts.

They can be found from the school yard to the prison yard. I am not trying to imply that arts and crafts is practiced solely by children and hardened criminals. It is also practiced by summer campers, college students, the unemployed, stamp collectors, homemakers, heartbreakers, earth-shakers, and grandmothers (I cannot emphasize the latter strongly enough).

Mr. Resistor Man Says: Allow the projects in this chapter to help you forget that there are approximately 6,000 new computer viruses released each month!

Mouse Pencil Sharpener, p. 146

As illustrated by the wide range of stereotypes just listed, both unexpected and not so unexpected people are practitioners of the arts and/or crafts. In fact, throughout much of known human history, people have been engaged in making arts and crafts. We know this because some of the earliest records of human history unearthed are of cave paintings and ceremonial craft objects.

Across cultures and throughout history, we silly humans have been using artistic creation as a means of self-expression. However, our artistic works tell people not only about ourselves, but also about who we are as a people. We may just be creating to better express ourselves or simply alleviate our boredom, but the artwork as an extension of our psyche, when framed

All-in-One Shadow Box, p. 148

Polarized Wallet, p. 151

Floppy Disk Notepad, p. 144

Squid-Skinned Camera Case, p. 157

Resistor Pillow, p. 166

in the context of the larger collective whole, provides a valuable tool for understanding a much larger social abstraction often referred to as culture.

It should not be much of a surprise, then, that in the heart of the digital age, in a book about repurposing broken electronics, there is a chapter dedicated to making arts and crafts. There is nothing, on a whole, that has captured our imaginations or shaped our worldview as much as the personal computer. It is only natural that we should use computers as the raw materials and tools to energize our creativity.

So put on your artist's cap, pull out your trusty screwdriver, and make something!

#33 FLOPPY DISK Notepad

LEVEL 1: Novice | **TECH TRASH:** 2 Floppy Disks

For me, the most important quality in a notebook is that it can fit in my pocket. A notebook made out of a 3.5" floppy disk can surely do that. That is what makes it so great! Well, that and the fact that you made it yourself and all of the materials are completely recycled. This is a quick and dirty way to make a clean little recycled notepad perfect for the Earth-conscious inventor on the go. In less than an hour, these little disks will be ready to store even more important information than before!

MATERIALS

▸ 11 sheets of scrap printer paper*
  *Used printer paper is great for notepads. Most people tend to print only on one side of the paper.
▸ Scissors
▸ Ruler
▸ Craft knife
▸ 2 floppy disks
▸ Hole punch
▸ Pencil
▸ Two 10" strips of wire (you can get these from a ribbon cable)

Mr. Resistor Man Says:
Computer Lib, written by Ted Nelson in 1974, is considered by most to be the first personal computing handbook.

MAKE IT

1. CUT YOUR PAPER

Take a sheet of scrap printer paper and cut two 3½" by 11" strips. Then cut three 3½" squares from those two strips. Repeat on about ten more sheets of paper (this will generate 60 pages in your notebook). Set them aside.

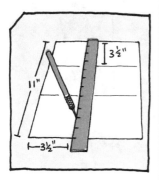

2. PREPARE THE FLOPPY

If your floppies are not already write-protected, pull down the nifty little "write protect" tab at the bottom of each disk. This will give you a second little hole through the bottom of the disk.

3. PUNCH HOLES

Center a page underneath one of the disks. Use a pencil to mark the paper below through each of the two holes. Remove the disk and punch holes at the two dots. Check that they're aligned, and punch more holes by aligning the paper over the remaining sheets.

4. HARVEST WIRE

Cut two 10" strips of wire from a ribbon cable. These will be used to bind your notebook.

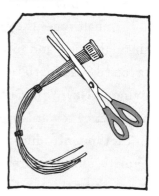

5. WIRE IT UP

Place one of the floppy disks facedown on your work surface. Stack the pages on top, aligning the holes. Insert the wire through one hole on one floppy disk and all of the corresponding holes in the stack of paper. Repeat with the holes on the opposite side of the notebook. Finish by inserting the wire ends through the two holes on the second floppy disk.

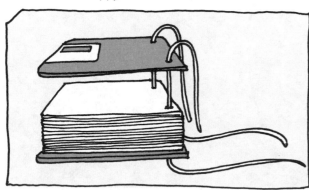

6. TIE IT UP

Tie the two ends of one wire in an overhand knot. The loop it creates should be large enough so the notebook can easily fold open but not so large that it falls apart. Repeat with the other wire, then trim the ends.

7. EVOLVE

Stop throwing your disks around like a monkey and start writing down your thoughts in your new notepad like a human being.

#34 Mouse Pencil Sharpener

LEVEL 1: Novice | **TECH TRASH:** Broken Trackball Mouse

Traditionally, mice are known to eat cheese at mealtime—they're often depicted burrowing through holes and nibbling on corners of their favorite Swiss variety. (And, as I discovered when living in New York City for a time, they also have a tendency toward large blocks of milk chocolate.) But let me introduce you to this little mouse—quite a rarity, since his preferred snack is pencils—the more blunt the tip, the better! He chews through wood and graphite with the voracious appetite of the most streetwise rodent. Result: Your pencil has an endlessly sharp point and your mouse isn't plagued by indigestion!

MATERIALS

- Screwdriver
- Broken trackball mouse
- Pencil
- Scrap wood
- Power drill with a ⅜" drill bit
- Hot-glue gun
- Metal pencil sharpener
- Epoxy
- Scissors
- Cardboard
- Diagonal cutters

Mr. Resistor Man Says:
A standard yellow #2 pencil contains enough graphite to write approximately 45,000 words.

What a "sharp" accessory for your desk!

MAKE IT

1. OPEN AND GUT MOUSE

Remove the screws from the underside of your dead mouse to open it up. Remove all of its electronic innards.

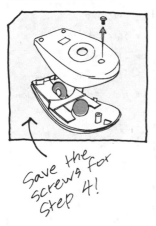

Save the screws for step 4!

2. DRILL

Flip the top part of the case over so it sits like a bowl and make a marking centered at the lowest point. Rest the case on a piece of scrap wood or a worktable that you do not care about and drill a hole through the plastic at the mark.

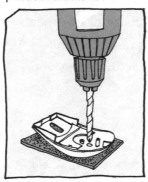

3. GLUE

Apply a small amount of glue around the pencil sharpener's circular opening and align it against the 3/8" hole in the mouse case. When the hot glue is dry and if you are happy with the alignment, make a more permanent bond by sealing around the edges with two-part epoxy.

4. TOUCH-UPS

If your mouse has a track wheel (used for scrolling), insert it in place in the bottom of the case. Apply glue around the inside edge of the opening in the top and snap closed the mouse case (to seal the opening). Replace any screws .

5. COVER HOLE

Cut out a 1" circle of cardboard (it doesn't have to be precise). Remove the piece of plastic from the trackball opening in the bottom of the mouse case, squeeze a small amount of hot glue over the inside surface, and press the cardboard circle over it. Then reattach the plastic piece, sealing the hole so pencil shavings don't leak out.

Squeeze a drop of hot glue into the hole in the front of the mouse case where the cable used to be attached. Let dry.

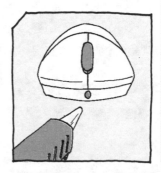

6. SHARPEN!

Gather up all your dull pencils and sharpen them to your heart's content.

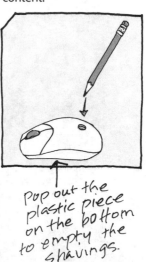

Pop out the plastic piece on the bottom to empty the shavings.

#35 All-in-one Shadow Box

LEVEL 1: Novice | **TECH TRASH:** All-in-one Printer/Copier/Scanner/Fax Machine

Let's face it, buying picture frames to display each and every photo you print can get pricey. One solution? Throw away all of your photos. The more sensible approach? Condense all of your memories to a single frame! A true marriage of form and function: This all-in-one-frame (for *all* your photo display needs) started out its life as an "all-in-one" printer/copier/scanner/fax machine. On account of its proven versatility in handling almost any image-processing task you sent its way, it should have little problem adding "the only picture frame you will ever need" to its repertoire.

MATERIALS

- All-in-one printer/copier/scanner/fax machine
- Screwdriver
- Diagonal cutters
- Strip of Styrofoam
- Colored construction paper
- Straight pins
- Scissors
- Craft glue
- Photographs (to cut up)
- Tape
- Wooden skewers

NICE MUFFS

See Page 107 to make these sweet Hearmuffs.

MAKE IT

1. REMOVE THE BED

Separate the scanner assembly from the rest of the machine. Use your diagonal cutters to cut any wires that prevent you from doing this.

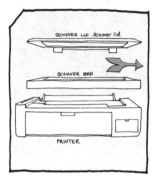

2. EMPTY IT

Take out the inside of the scanner assembly and set aside the parts.

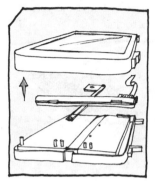

SAFETY FIRST

>> **Be careful not to break the scanner's thin fluorescent bulb. Remove and recycle it appropriately.**

3. CUT TO FIT

Take the foam and cut it into a strip to fit in the edge of the scanner tray.

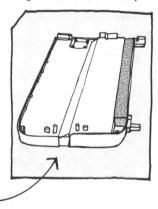

The foam strip will be your "planter," in step 8.

4. COVER THE FOAM

Wrap a piece of black construction paper around the foam strip to hide it. Hold the paper in place with straight pins.

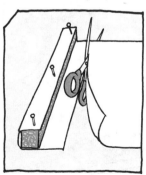

5. BACKGROUND

Cut shapes in different colors of construction paper. Arrange and paste them on a full sheet of construction paper with craft glue to create a background image. Get creative!

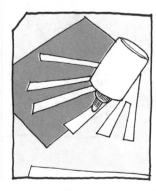

6. CUT OUT

Cut out the subjects (people, pets, odd architecture) from your photographs. Also cut out shapes from the remaining construction paper to use as foreground decoration. Set these aside.

continued >>

7. SKEWERS

Tape the end of a skewer flat to the back of each of your photo cutouts. Trim the skewers so they fit comfortably inside the "frame" of your scanner assembly.

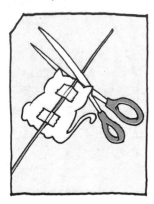

8. IMPALE

"Plant" the bottom of the skewers into the foam. Arrange all of the images to your liking.

9. FOREGROUND

Insert shapes you cut in Step 6 into the foreground to add extra decor to the scene and to hide the skewers and foam base. Arrange all the layers to fit.

HELPFUL HINT

Don't go too overboard with foreground decor or it might obscure the actual photos. A little bit of choppy grass or a few shrubs should do the trick.

10. CASE CLOSED

Close the casing for the scanner assembly, replacing any screws.

11. STAND

Stand your frame upright. Reattach the scanner cover by inserting it backward into the frame to create an L shape, as shown in profile.

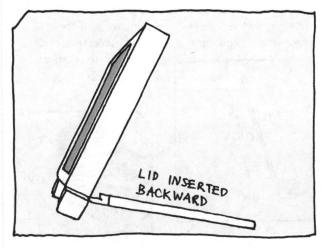

LID INSERTED BACKWARD

#36 Polarized Wallet

LEVEL 2: Intermediate | **TECH TRASH:** LCD Monitor or Laptop with Broken Backlight

The beauty of this wallet is that depending on how the lighting hits it, sometimes people can see you have money and sometimes they can't. This allows you to be as broke or rich as you need to be. Like the glass half empty/glass half full model, it is all a matter of perspective. For instance, when you're out with your buddies and need to chip in for pizza, you can appear to be dead broke, but later on, when you are out on a date, you can appear to be filthy rich! Of course, you wouldn't actually do that, would you?

> ### MATERIALS
> - LCD monitor or laptop with a broken backlight
> - 8" by 3" black-and-white image printed on transparency (this can be done at your local copy shop)
> - Screwdriver
> - Ruler
> - Scissors
> - Clear packing tape

> **Mr. Resistor Man Says:**
> Iceland spar, a naturally occurring transparent crystal, was key in the scientific discovery of polarization by Erasmus Barthonlinus in 1669.

MAKE IT

1. EXTRACTION

Remove the LCD screen assembly from the laptop by taking out the screws that fasten shut the plastic frame around the screen.

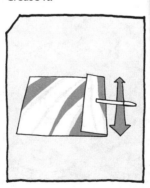

2. DISASSEMBLY

Take apart the LCD assembly and remove the polarizing filter. It is the thin sheet of plastic located between the actual LCD module and the backlight diffuser sheet.

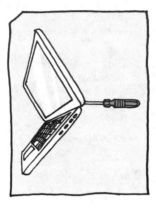

POLARIZING FILTER

HELPFUL HINT

Identify the polarizing filter by its translucency and weird optical properties—it lets light through in strange ways and may multiply or magnify things placed on the other side of it.

3. START FOLDING

Turn the plastic horizontally and make a vertical fold so that you have an even crease 3" in from the right edge. Crease it.

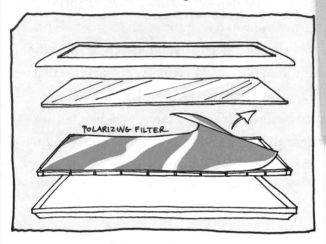

HELPFUL HINT

To make a nice sharp crease in this material, you will have to press down firmly upon the crease and move your finger back and forth a number of times along the crease. You can also use the blunt, round end of a screwdriver instead of your finger.

4. KEEP FOLDING

Measure 3" in from the fold at the right edge of the page. At the 3" mark, fold the folded edge from Step 3 underneath and make a sharp crease. You should have folds that look like an exaggerated Z.

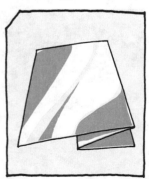

5. AND FOLD AGAIN

This last major fold will create the card sleeve. Flip your sheet over so that the shorter side of the Z is on top. Fold the excess from the bottom sheet over onto the top layer. There should now be a money fold and a card fold.

6. CUT NOTCHES

Cut a small ¼" notch in the creases of the four corners of the outermost layers.

7. CLOSE THE ENDS

Take the two sheets of plastic created by the notches cut in Step 6 and fold them in toward each other, making creases where the notches stop. To secure the folds, tape each little flap to the larger sheet (innermost plastic sheets) that it now touches. Repeat on the other side.

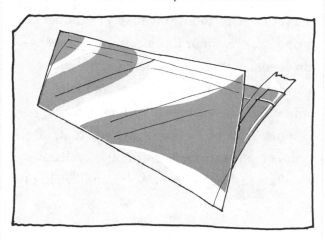

8. TRIM

Trace and trim the transparency so it's about ⅛" smaller than the open wallet. Insert it into the back pocket. It should fit snugly and butt up against the pocket closing you folded in Step 7.

9. REPEAT

Fold the remaining open side of the pocket closed just like you did with the other side in Step 7.

10. TAPE IT

The excess plastic on each side of the back flap should now be folded toward the card sleeve. After this fold is made, apply tape to hold this plastic in place. Fold the wallet in half.

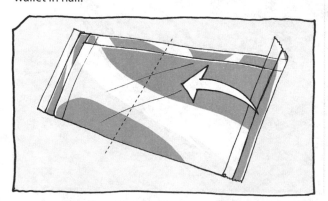

11. SLICE IT

Cut a slit all the way down the middle of the card sleeve (cutting it in half). Trim off the inside corners to round them.

12. FILL IT

Fill your wallet with money and credit cards. If you don't have one already, this may require getting a job.

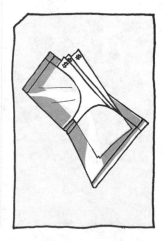

#37 MY FIRST cell Phone

LEVEL 1: Novice | **TECH TRASH:** Cell Phone

Discounting the cell phones that were lost and stolen, and that one that ended up on the bottom of a canal (don't ask), I still have every single cell phone I've ever used, including my first. I've thought for some time about the proper way to honor this electronic pioneer in my life, for it just wouldn't seem quite right to merely scrap such a precedent-setting phone for a handful of parts. Such a landmark device is worthy of commemoration. It deserves to be honored and admired by all those yet to come. That is why I have resolved that, after removing all of the electronic components from the phone, I will have it bronzed to display on my office desk as a paperweight for all visitors to gaze upon in reverence.

MATERIALS

- Your first cell phone (or, really, any cell phone)
- Torx wrench
- String
- Mixing stick
- Two disposable mixing bowls
- Plaster
- Paper towels
- Chair
- Natural-fiber brush
- Bronze powder
- Newspaper
- Spar or marine varnish
- 220-grit sandpaper
- Scissors

MAKE IT

1. DISASSEMBLE

Disassemble the cell phone. Set aside the parts you want to reuse, such as the tiny vibrating motor and the microphone, and recycle the rest, like the screen and battery.

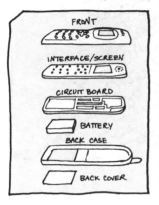

2. STRING IT UP

Tie a string about 2' long to the bottom inside of the phone (it needs to be long enough to be able to hang the phone upside down from a chair).

3. STIR IT UP

With your mixing stick and a small plastic bowl, stir up the plaster as directed on the packaging.

4. PLASTERED

Position the phone pieces in the manner that you want them preserved and then carefully pour the plaster into the two halves of the phone. If you pour too much and it gets on the outside of the case, wipe away the excess with a damp towel.

5. REASSEMBLE

Before the plaster is fully dry (it should start to have the consistency of clay or mud), reassemble the cell phone case. Clean the excess plaster off the outside of the case with a wet paper towel.

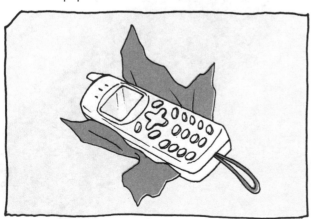

6. HANG

Hang the phone (attached to the string) outside from the underside of a chair or other solid structure.

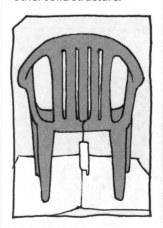

7. MIX

Using your brush, mix the varnish with a few small scoops of brass powder to get a nice solid bronze color. Continue stirring until all the brass clumps have disappeared.

continued >>

8. APPLY

Place a newspaper under your work area to protect the ground from getting stained bronze. Apply a thick coat of the bronze varnish to all surfaces on the phone.

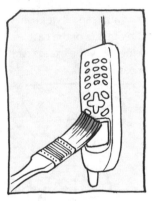

9. EXPECTED WAIT TIME

Wait at least four hours for the varnish to dry (or as directed).

*Start another project during the downtime while you wait.

10. ANOTHER COAT

Once the varnish is dry, lightly sand the phone with your 220-grit sandpaper and then apply another coat of varnish. Repeat this process of painting and waiting until you are pleased with how the phone looks. Once satisfied, don't sand the phone after the last coat.

SANDPAPER

11. CUT IT DOWN

Cut the phone off the string. Carefully coat the spot where the string was with the brass varnish. Let dry.

12. DISPLAY

Proudly display the testament to your personal and technological growth on your desk as a paperweight.

#38 SQUID-skinned CAMERA CASE

LEVEL 2: Intermediate | **TECH TRASH:** Keyboard

Once, on a visit to the American Museum of Natural History, I came to believe that giant squid fought at the depths of the ocean with whales in a similar manner to Godzilla battling Megalon. Yes, I was one of those quiet kids with an overactive imagination. My parents always just thought that my curiosity and rampant imagination was a phase that I would grow out of. But, 20-some-odd years later, I still think it would be awesome to reach into my backpack and find a giant squid tentacle reaching out to pull me in. I'm not sure that everyone is going to appreciate this project on the same level as I do, but at the very least, it will be very easy to find your camera amidst all the other gear in the darkest of backpacks (should you have the courage).

MATERIALS

- Keyboard
- Phillips-head screwdriver
- Ruler
- Marker
- Scissors
- White cotton flannel or similar (at least 6" by 14")
- Straight pins
- Sewing needle and thread

MOM?

MAKE IT

1. OPEN THE KEYBOARD

Use a Phillips-head screwdriver to open the keyboard.

2. REMOVE THE RUBBER

Remove the rubbery plastic sheet between the keys (it looks like a squid tentacle) and the thin plastic circuit sheeting.

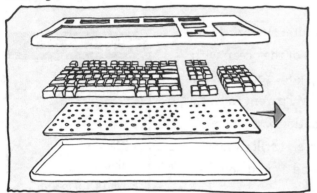

3. MEASURE YOUR CAMERA

Measure the length, width, and thickness of your camera. Write it down so that you don't forget. Multiply the length by 2 to get x. Add the width to the thickness and add ½" to get y. For example: If your camera is 4" long by 2¼" wide by ¾" thick, x equals 8 and y equals 3½".

4. CUT THE RUBBER

Mark and cut a rectangle measuring x" by y" from the rubbery plastic. Then cut a rectangle ¼" longer from the fabric.

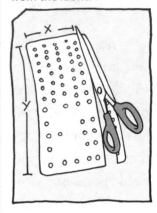

5. FOLD

Center the rubber piece on the fabric, tentacle bumps facing up. Mark the length measurement of the camera along one of the long edges. At this mark, fold both layers together so the tentacle bumps face inward.

6. PIN

Pin the corner edges of the rubber piece and soft fabric together to keep the fold.

Don't forget to remove the pins as you start to advance with your sewing!

7. SEW

Start at one folded corner and sew a whipstitch along the edge, stopping ⅛" before the end of the fold of soft fabric. Fold the ⅛" lip back, curling both layers of material over each other and pinch it in place. Continue sewing along the lip's edge, securing it down.

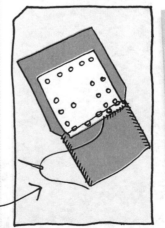

HELPFUL HINT

If you have a sewing machine, you can use it to complete this step much faster and with stronger stitching.

8. FLIP OUT

Turn the pocket right side out (the rubbery tentacles should now be on the outside.)

9. SEW YOUR LIP SHUT

Fold the rubber lip that is already partially sewn toward the inside of the pocket. Sew it in place using a running stitch.

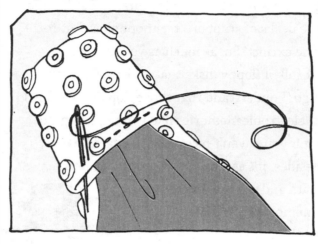

HELPFUL HINT

Running Stitch: Insert your needle and pull the thread toward the inside of the case and, once through, flip the needle around and pull the thread back through toward the outside of the case.

10. SEW THE SIDES OF THE FLAP

Tuck in the extra fabric under the flap and sew a whipstitch along the sides, closing them up.

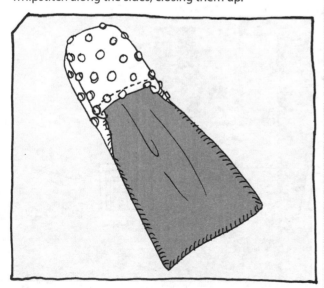

11. SEW THE TOP OF THE FLAP

Sew the top of the flap with a whipstitch, the way you sewed the sides.

12. CLEAN IT UP

Trim all the thread ends and slip your camera in!
Optional: Add a small hole for the camera strap. Remember to sew around the edges of the hole to keep it from ripping apart.

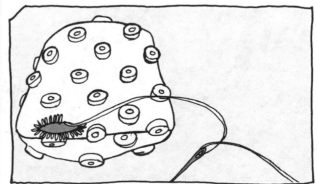

#39 FLOPPY DISK Wall Frame

LEVEL 1: Novice | **TECH TRASH:** Floppy Disks

Floppy drives have gone the way of the buffalo. Sure, you still might find a few roaming the desks of stubborn technophobes, but for the most part, they've gone extinct. So, as much as you may want to hold onto that shoebox full of floppy disks that is sitting in your basement, it's time to let go. I understand that those floppies are filled with countless wonderful electronic memories like your eighth-grade paper about oak trees, but if you haven't recovered the data by now, you are never going to. Besides, it's essentially irretrievable. You may as well forget about the data and use the disks to display more important memories: photos—the kind of memory that is instantly retrievable. Unless, of course, like your floppies, you've got those buried somewhere deep in your basement, too.

MATERIALS

- Foam board
- Handful (10 to 20) unwanted floppy disks
- Hot-glue gun
- Metal ruler
- Craft knife
- Hammer
- Two to six 1" brads (small nails)
- Photos!

Mr. Resistor Man Says:
What a waste! Floppy disk drives were built without serviceable parts because manufacturers felt that buying a brand-new drive would be considerably cheaper than the hourly cost of repairing one.

MAKE IT

1. PLAN

Lay your foam board flat across your work space. Arrange your floppy disks on the boards in a slightly staggered brick pattern so they are all touching but not in a perfect grid.

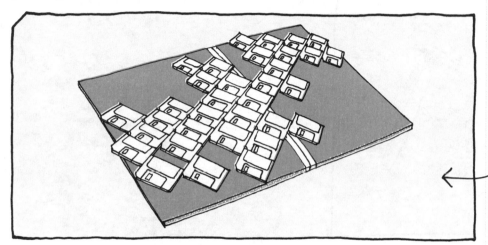

If you are using more than one sheet of foam board, arrange the disks diagonally along the length of the boards to better distribute their surface area over the seam.

2. GLUE

Once you have a pattern that you like, glue down the floppy disks. Apply hot glue liberally so that it covers the back of the floppy disk leaving a ½" allowance at the edges.

3. CUT

Use your craft knife to cut around the outside of the floppy disk shape to remove the excess foam board. Then flip the project over and cut ½" off the edges of the foam board around the entire perimeter (shown).

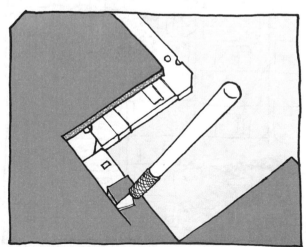

4. NAIL IT

Nail the frame to the wall by inserting several 1" brads through the foam in the small spaces between the disks.

5. PICTURE TIME

Hang pictures and postcards in the frame by sliding them behind the metal tabs on the disks.

FLOPPY DISK WALL FRAME

>> Variation

★ Rather than one big frame, arrange the disks in several small Tetris-shaped frames.

★ Glue a magnet to the back of a disk to make a refrigerator-friendly frame.

★ Line up the discs in a grid to create a large wall calendar!

#40 OBLIGATORY decorative candleholder

LEVEL 1: Novice | **TECH TRASH:** Hard Drive

Every holiday season I receive some sort of decorative candleholder from someone I know. There can only be two reasons for this: Either I seem like the kind of guy who likes decorative candleholders or other people must really like them. After having met myself once or twice, I don't think I am the type of guy who likes candleholders and it's probably more likely that most everyone else does. With this in mind, I am going to show you how to make a decorative candleholder because you are probably in the large section of the population that glows bright just thinking about new ways to display candles. If this project isn't for you, make it and give it away as a gift when the holidays roll around!

Mr. Resistor Man Says:
Did you know? LEDs are measured in millicandela (mcd), which are 1/1000th of a candela (the light intensity of a single candle). That means that an LED is 8,000 mcd is as bright as eight candles.

MATERIALS

- Phillips-head screwdriver
- Torx wrench
- Old computer
- Hard drive
- Pliers
- Large, colorful, scented candle

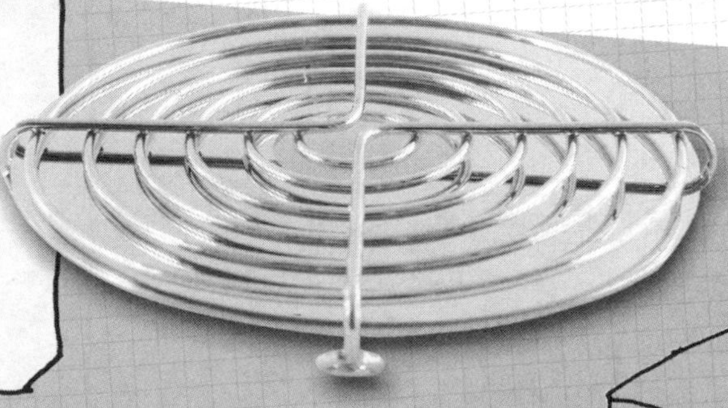

MAKE IT

1. OPEN THE HARD DRIVE

Use your screwdriver and/ or Torx wrench to carefully remove the cover from the hard drive. (This, will void the warranty. Your hard drive is officially dead now.)

Skip Ahead

If you have already made a Magnetic Actuator Arm Picture Display (see page 60), then you should have some extra hard drive platters lying around. You can skip to Step 3!

2. REMOVE THE HARD DRIVE PLATTERS

There should be a series of screws clamping the platters (shiny discs in the hard drive) to the spindle. Remove the screws and the plate that is clamping them down. Rotate the actuator arm so that it is no longer sandwiched between the platters. Finally, remove the nice shiny platters.

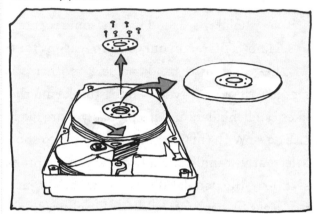

3. FIND THE FAN GUARD

Locate the fan guard in the back of the computer case or attached to the fan *inside* the computer's power supply. Unscrew it.

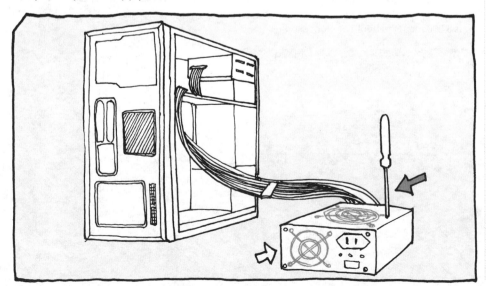

*An average PC should have roughly two fan guards

4. START THE TRANSFORMATION

Decide which surface of the fan guard you want to be on top and, using your pliers, bend all of the "legs" of the fan guard back (away from the "top") about 45 degrees.

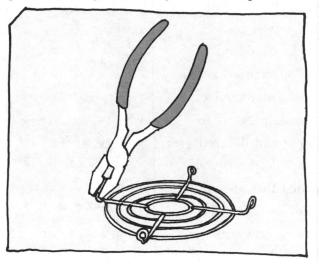

5. STACK THEM UP

Place the fan guard facedown and center the hard drive platter on top of it. Slowly and carefully bend the legs of the fan guard over the hard drive platter until they are bent inward enough to hold the platter firmly in place.

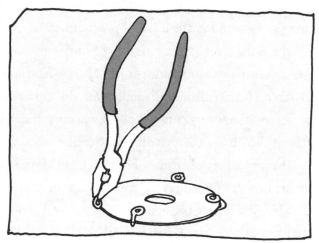

6. LIGHT THE FLAME

Flip the assembly over. Your candle holder should stand flat on all four legs. If it's a little wobbly, carefully adjust the legs until it is level. Then place a large, colorful, vanilla-scented candle on top!

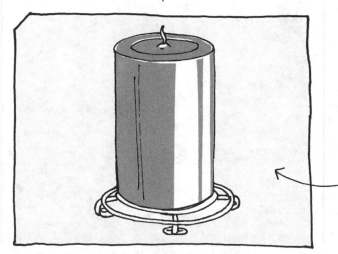

It doesn't have to be vanilla! For a relaxing evening, try lavender—or rose for something romantic.

#41 Resistor Pillow

LEVEL 4: Geek Squad | **TECH TRASH:** Circuit Board

This project came to me in a mesmerizing dream of circuit boards and wizards. I awoke suddenly from my slumber, waved my imaginary wand in the air, and proclaimed, "Resistor pillow!" Okay, maybe not. In reality, my nightly sojourns are usually too strange to wake me up with any actionable project ideas—a typical dream may involve anything from running through walls during a rugby match at work to being eaten alive by cute wolflike creatures that are all singing different television sitcom themes off-key. Needless to say, I don't sleep very well.

This project really came about because I love resistors! They are one of the most ubiquitous electronic components and they are especially durable. If a device breaks, the resistors will almost always still work. It is thanks to their familiarity and reliability that I find resistors particularly comforting, and hopefully a giant functioning resistor pillow will help with naps! After all, getting more sleep should keep those neurons in my head from short-circuiting and causing nightmares. Neat-o!

Mr. Resistor Man Says:
Don't be intimidated by the size of the materials list. It's really just some fabric, thread, and glue. There isn't much here a typical crafter won't have on hand.

MATERIALS

- Almost any circuit board (from a scanner, printer, or sound card)
- Wire cutters
- Sheet of 18" by 24" paper
- Pencil
- Scissors
- Ruler
- Yard of beige fabric
- 1' each of fabric for each color stripe on the resistor
- Straight pins
- Sewing needles and thread to match all fabric
- 1' by 5' strip of conductive fabric
- String
- 19-gauge stainless-steel wire
- Gaffer's tape (or duct tape)
- Polyester fiberfill batting
- Soldering iron and solder
- Hot-glue gun
- Pliers
- Crimp-on ring tongues (16-gauge)
- Two 3/16" eyelets and eyelet tool
- Stranded hookup wire
- Two 6-32 by 1/2" nuts and bolts
- Wire coat hanger (optional)
- Phillips-head screwdriver

MAKE IT

1. FIND A RESISTOR

Examine your circuit board and find a resistor you like. Cut it from the board with your wire cutters, leaving as much wire attached to the resistor as possible.

If you are curious, you can determine the value of your resistor by looking it up on the table in Chapter 1, page 11. The resistor that I used is a 10K (10 kilo ohm) resistor.

2. MAKE THE SHAPE

Take a piece of paper and fold it in half lengthwise and then widthwise so you are left with a folded piece of paper a quarter of the size of the original sheet. Turn your paper so that it appears to be oriented as a "landscape," with one crease on its left side and one crease on the bottom of the paper nearest you. Draw an arc from the crease on the bottom to the crease on the side, as shown, mimicking the top quarter part of a resistor. Draw a second line ½" outside the line you just drew.

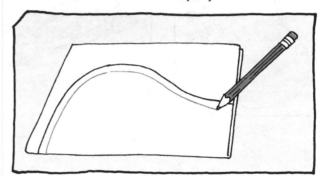

3. CUT OUT THE SHAPE

With the paper still folded, cut along the outer marked line. Then unfold the paper to reveal the shape of the resistor.

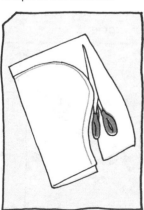

4. DRAW THE PATTERN

Measure and draw four evenly spaced 2" stripes on your pattern to represent the stripes on the resistor.

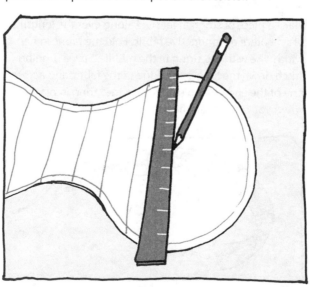

5. TRACE AND CUT

Trace the pattern onto your beige fabric twice and cut out the two identical pieces.

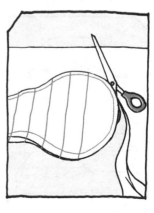

continued >>

6. MARK STRIPES

Line up the pattern over both fabric pieces. Make ½" snips into the fabric at the end of each marked stripe on the pattern to use as guidelines.

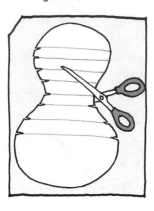

7. CUT STRIPES

Lay the 1' long fabric pieces flat. Measure the pillow and then mark and cut stripes that are 3" wide and long enough to span the length of one side of the pillow. You will need to cut two matching stripes for each color (one for each side).

8. "HEM" STRIPES

Lay the fabric strips flat, wrong side up, and fold the long edges in ½". Pin them in place.

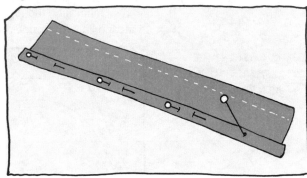

9. PIN STRIPES

Pin the stripes right side up, as shown, onto the pillow fabric to hold the stripes in position while you sew.

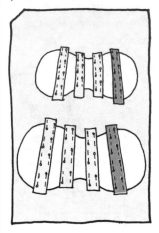

10. SEW STRIPES

Sew the stripes onto the pillow fabric pieces using a running stitch. Remove the pins and set aside the pillow pieces.

11. CUT LEADS

Mark and cut two 4" by 42" strips of conductive fabric for the resistor leads.

12. SEW LEADS

Place a piece of string at least 44"-long over the length of one piece of conductive fabric. Fold the fabric in half widthwise with the string in the middle. Sew a running stitch down the long open edge of the fabric and across one of the ends, being sure to sew the string in place when you sew closed the end .

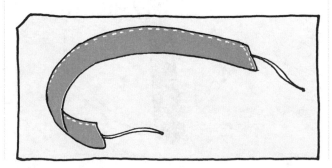

13. INSIDE OUT

Tie the string to a doorknob or a table leg (or have a friend hold the end) and pull gently on the fabric to turn the tube inside out. Then cut the string away. Set the tube aside to use as your plush resistor lead.

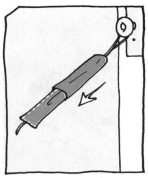

14. WIRED

Take a piece of stainless-steel wire that is approximately 3" longer than your lead and round one end into a hook so that it won't push through your fabric. Wrap a piece of gaffer's or duct tape to secure the loop. Insert this into the long tube of the resistor lead. Repeat the Steps 12 to 14 to make the second lead.

15. START STUFFING

Stuff the resistor leads with batting until they are firm but not too rigid. Trim flush any excess wire left sticking out of the end.

HELPFUL HINT

Use a blunt (folded) end of a coat hanger to push the batting in (the rounded end won't puncture the fabric).

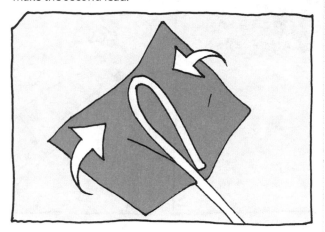

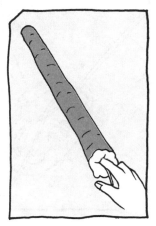

16. PIN IT AGAIN

Pin together the two pillow pieces from Step 10, right sides in, with the open ends of the resistor leads tucked into the ends of the pillow. Leave an opening about 5" wide along one of the long sides of the pillow for turning it right side out. The ends should stick out about 2" to 3". (The rest of each resistor lead should be tucked inside or arranged so they come out of the opening on the side.)

17. SEW

Use a running stitch to sew around the pinned edges of the pillow, leaving the opening. Be careful to only stitch the resistor leads into the seam on the ends where you are sewing them together with the pillow.

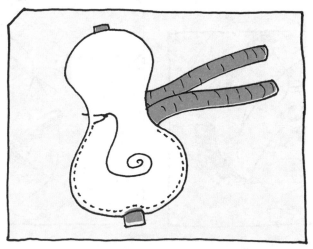

18. FLIP OUT

Gently pull the leads through the opening to turn the resistor right side out. Set the pillow aside.

* Take your time turning the pillow inside out.

continued >>

19. SOLDER

Solder two 2' long wires to the actual resistor.

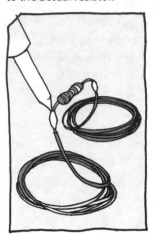

20. REINFORCE

Being careful not to break the solder connection, tie the wires together in an overhand knot with the resistor in the middle. Encase the knot in hot glue.

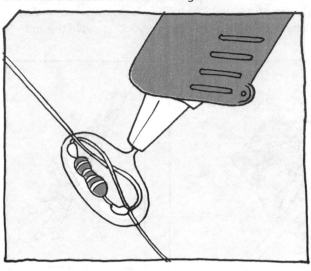

21. CONNECTORS

Strip ½" of jacketing off the end of each wire. Slide the ring tongue over the unjacketed wire and crimp it in place with a pair of pliers.

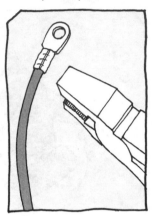

22. ROUND

Pull the stubby end of the stuffed resistor leads out from the inside of the pillow (these are the 2" to 3" ends not sewn shut). Bend the ends of the steel wire that is inside the resistor lead and wrap them with gaffer's tape to prevent them from stabbing through the pillow (like you did with the other end in Step 14).

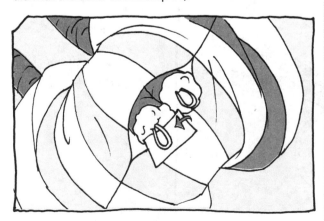

23. EYELETS

Poke a hole through both fabric layers at one end of the stuffed resistor leads. Widen the hole to be large enough for an eyelet. Insert an eyelet and crimp it in place. Repeat with the other lead.

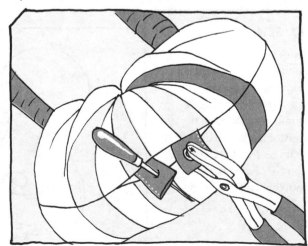

24. STUFF HALFWAY

Stuff the bottom half of the pillow with batting.

25. BOLT

Fasten one of the ring tongues to one of the eyelets using the matching 6-32 nut and bolt. Repeat on the other end. Your pillow should now function as an actual resistor.

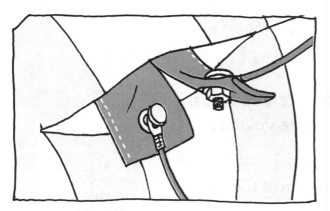

26. SEW CLOSE

Finish stuffing the resistor pillow with batting. At the opening, fold the lips of the fabric inward and use a whipstitch to sew it shut. When you get to the point that you can no longer fit your fingers inside the pillow, sew the remaining hole shut using slipstitches—or what I like to call the Frankenstein stitches (see helpful hint, below left).

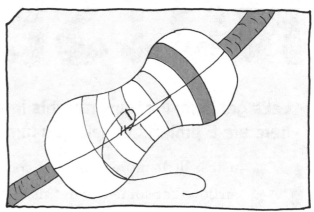

HELPFUL HINT

Slipstitches are small X's that look like the stitching on Frankenstein's forehead. To make them, sew along the outside of the seam in one direction and then back in the other, forming tiny X's.

CHAPTER 6

Making
NOISE

> **Let's get loud: From instruments to speakers to amps, here are 6 projects to help you turn up the volume.**

Music is little more than an orchestrated disturbance of airflow. Yet somehow in this disruption of the aether, we perceive something greater, something ineffable. In this variation we find harmony and even transcendence, inspiring people throughout the ages to listen to, dance to, and making music.

Alright, perhaps I am laying it on a little thick there, but when you give it some thought, it's hard to perceive music as anything short of magic. After all, it is a great challenge to fathom how this impalpable movement of air so completely enchants. And yet,

Mr. Resistor Man Says:
Beyond dead electronics, you can make music with your tools, too. With a cello bow, you can learn to play a handsaw with a flexible blade.

PROJECTS

Gamer Guitar, p. 193

music is all around us. It is hiding in tensioned metals, stomping up stairs, and it is drip, drip, dripping out of leaky faucets. There is not a thing on Earth that doesn't make or can be used in some capacity to make music.

Just look at a pile of dead computers and see all the music just trying to escape. In some instances it will be quite obvious. Some parts of a broken computer are already suited to make music. An old plastic computer casing can become a percussive drum; and there are plenty of metal sheets that can be pried from the inside to make some oddly shaped alterna-cymbals.

In other cases it requires a little more coaxing and imagination to release an object's secret music. That old mouse in the junk drawer requires a bit of transformation to be come the perfect shaker.

Mouseracas, p. 181

Acoustic Rock Revolution, p. 184

Portable Amplifier, p. 188

Music Monster, p. 176

A broken gaming system doesn't exactly scream out to become a rocking guitar (at least not at first!). Rather, you need to look at that console and imagine that a few extra parts and a set of strings could make music resonate through its hollow chamber.

This chapter is by no means the extent to which broken computers can be used to make music, but it is a brief glimpse into a vast world of possibility. We are barely touching the surface of an area as infinite as the possibility of fluctuated air currents themselves. So dig in, figure out how things work, and let these next handful of projects be your entrée into this computer based orchestra of the absurd.

#42 Earbud SPEAKERS

LEVEL 1: Novice | **TECH TRASH:** Earbuds

If you're anything like me, you have gone through so many MP3 players that you have accumulated a pile of extra, unused earbuds. They're not exactly dead, but at the same time, it's highly unlikely that you have acquired any more ears during this time period to accommodate all of these extra musical earplugs. Oh, what to do? Take those unused earbuds, mix in the magical power of acoustics, and make exceedingly awesome speakers, of course.

MATERIALS

- Working pair of earbuds
- Two nonworking MP3 players (for the base)
- Thin knife
- Mini screwdriver set
- Pliers
- Two large plastic funnels (with holding tabs)
- Two 10-24 by ½" bolts
- Two 10-24 nuts
- Hot-glue gun

Mr. Resistor Man Says:
As with most speakers, the tiny ones inside earbuds are transducers and can also be used as microphones.

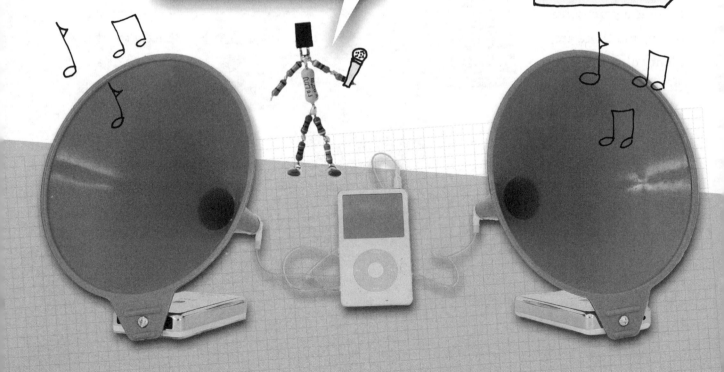

MAKE IT

1. OPEN IT

Open up an MP3 player. If you are using an iPod, this involves cautiously wedging a pointy knife into the side of the case and prying it open.

2. REMOVE SOME STUFF

First remove the audio jack, because you will be mounting the funnel to the case through the audio port. Then remove the battery (for safety reasons, since batteries can leak over time). Everything else should be left in the case because it adds weight.

3. FASTEN

Fasten the funnel to the bottom of the case with a nut and bolt so that the nut is on the inside of the case, going through the audio port.

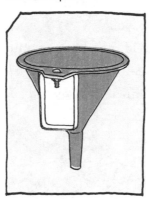

HELPFUL HINT

If your funnel did not come with a hole already in the holding tab, drill (or cut) one just large enough to accommodate the bolt.

4. CLOSE IT

Close the MP3 player case back up. The funnel should now be standing on edge, thanks to its new base. Repeat with the second MP3 player and funnel.

Funnels come in fun colors, so don't be afraid to get flashy.

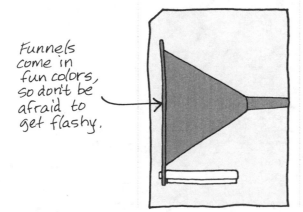

5. GLUE

Glue an earbud, speaker first, to the center of each funnel spout.

6. PLUG IN

Plug your newly constructed speakers into a working MP3 player and turn up the volume.

#43 Music MONSTER

LEVEL 3: Expert | **TECH TRASH:** CD-ROM Drive

Unless you live in the Magical Land of Make-Believe, that CD-ROM drive you've pulled out of your old computer is monstrously ugly and downright horrific. But I'd be willing to bet that it's probably still perfectly functional as a CD player.

So in keeping this horrid metal box around to use as a CD player, let's *embrace* its lack of beauty. Nay, let's *revel* in its grotesqueness and transform this drive into the *biggest, baddest* music monster you can think of.

MATERIALS

- Small machine screw and corresponding nut
- 7805 voltage regulator
- Heat sink
- Computer drive power connector
- Wire stripper
- 12V 500mA power adapter
- Multimeter
- Marker
- Tracing paper
- Soldering iron setup
- Coat hanger
- Ruler
- Zip ties
- Nonamplified computer speakers
- Hot-glue gun
- CD-ROM drive (with a play button in the front)
- Craft knife
- Scissors
- Hairy fabric (a cheap, furry bath mat can work!)
- White felt

Mr. Resistor Man Says:
The first CD ever burned was Richard Strauss' *Eine Alpensinfonie* (An Alpine Symphony) performed by the Berlin Philharmonic in 1983.

MAKE IT

1. HEAT SINK

With a machine screw and nut, fasten the voltage regulator to the heat sink so the flat side of the heat sink is flush with the voltage regulator and the voltage regulator pins stick off the bottom of the heat sink.

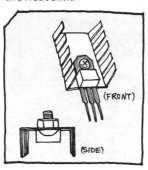

2. GET BENT

Prepare the voltage regulator by carefully bending the two outer legs 90 degrees to the side and the other 90 degrees away from the center (out to the front or back).

3. PREPARE THE PLUG

Prepare the power adapter by cutting the connector off the end and splitting the wire in half, using the indentation in the middle as a guide. This will create two separate wires on the end of the power cord (one will be the positive voltage and one will be ground).

4. STRIP THE WIRE

Strip the jacketing from the end of each wire. Plug the phone charger into the wall, and with your multimeter, check to see which is power and ground. Mark the ground wire with your marker.

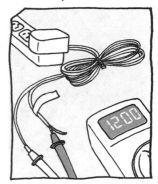

5. WITH A TWIST

Strip the ends off all the wires on the drive connector. Twist the two black wires with the ground wire from the 12V transformer. Twist the positive 12V wire from the transformer with the yellow wire on the drive connector.

The voltage regulator converts 12V to 5V, which is important, because the CD-ROM drive requires both a 12V and 5V power supply. This makes the voltage regulator perfect, since it is capable of providing both voltages.

6 SOLDER

Look at the voltage regulator head-on with the pins pointed down. Solder the ground wires to the center pin. Solder the yellow wires to the pin on the left. Solder the lone red wire to the pin on the right.

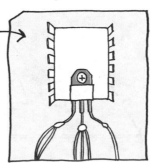

Extra credit!

Some old cell phone plugs have both 12V and 5V terminals. If you happen to have one, you can use this in place of the voltage regulator. Connect ground to the two black wires, 12V to the yellow wire, and 5V to the red wire.

continued >>

7. TAILBONE

Take a piece of thick, solid coat hanger wire and cut it so that it is 1" shorter than the wires in your power connector. Curl the ends to form loops.

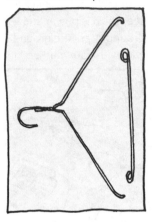

8. ATTACH

Pass the wires on one end of the connector through one of the loops in the coat hanger wire. Then twist the power wires around the coat hanger until you reach the second loop. Pass the ends through this loop.

9. CLAMP

Clamp the loops shut and gather all the wires together by fastening them with zip ties. (Neat wires prevent short-circuits!) Set them aside.

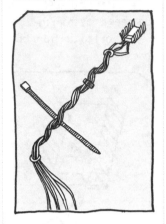

10. DISASSEMBLE

Completely disassemble the speakers so that you are left only with two speaker elements connected to an audio cable. Set them aside.

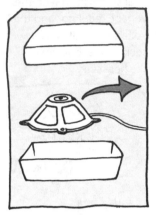

11. PYRAMIDAL

Retrieve two solid pieces of the speaker casings that are about the same size. Angle them to form a pyramid atop your CD-ROM drive and glue the edges in place with a hot-glue gun.

12. TRACE

Trace the front panel of the CD-ROM drive, marking the location of the CD tray, buttons, and audio jack.

13. CREATE A STENCIL

With a craft knife, cut out the markings in your tracing where the tray, audio jack, and buttons are. Also, cut out the outline of the front border, leaving an extra ½" on all sides.

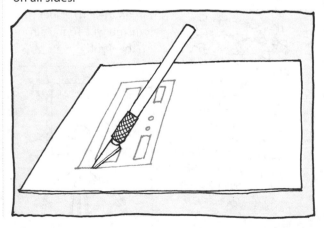

14. FURRY PANEL COVER

Flip the fabric furry side down and place your stencil backward on the fabric to trace the front panel. Cut out the furry panel cover with your craft knife.

15. GET FUZZY

Cut out a 2' square of furry fabric. Glue one edge of the square to the front top edge of the case.

16. EYE SOCKETS

Pull the furry fabric back over the pyramid. Figure out where on this mound you want the speaker eyes to be and then mark or poke holes in the back of the fabric to mark the spot. Flip the fabric over so the backside is facing up. Cut larger slits in the fabric where the eye holes are until it is large enough to pass the speakers through.

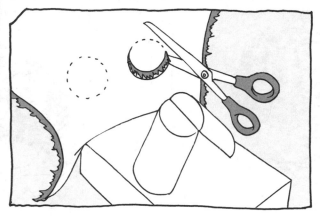

17. EYES

Pass the speakers through the holes and then glue them in place on the backside of the fabric. Then, take the speaker wire and run it through the pyramid and off the back of the drive.

18. SKIN IT

Pull the fabric taut so that the eyes are aligned and then glue the center of the fabric to the top of the pyramid.

19. FACE TIME

Before gluing, trim a little fur from around the holes for the buttons. Carefully glue the furry panel cover to the front of the drive. Do not glue the drive shut or glue down any buttons!

20. A LITTLE OFF THE SIDES

Starting from the front and slowly working your way back, continue to trim and fluff the furry fabric to your liking and glue the backing to the sides of the case.

Give your monster a hair cut!

21. REAR-ENDED

Check to make certain that the audio wire is extended fully back, away from the case. Then trim and glue the remainder of the fabric to the top edge of the back of the CD-ROM case.

continued >>

22. PLUG IT

Plug in the power connector to the back of your monster and bend its tail in such a way that the heat sink is freestanding and away from the case. It is important that the heat sink not be touching or leaning against anything, since it will get very hot when the monster plays music.

23. EJECT

Plug in the power cable and hit the eject button to release the CD tray.

24. DENTISTRY

Draw and cut a row of jagged teeth out of a piece of white felt to cover the length of the front of the CD tray. Then glue the row of teeth to the front of the disk tray. Once it's dry, you can load the tray back into the player.

25. LET THERE BE NOISE!

Put the audio plug into the audio jack in the front of the case. Then insert a CD and rock out with your furry little Music Monster.

SAFETY FIRST

>> Like most monsters, this Music Monster has a short temper and sometimes overheats. To keep your Music Monster from getting too hot and bothered about having to regulate his own voltage, always unplug him when you are done.

#44 Mouseracas

LEVEL 1: Novice | **TECH TRASH:** Trackball Mouse

When it was still part of my computer setup, I used to shake my trackball mouse all the time. I wasn't necessarily doing it in a particularly rhythmic way to try to make music. Rather, I was trying to get the darned thing to work! Needless to say, shaking your mouse around and banging it against the desk turns out to be the improper way to fix it. In fact, it's a highly effective way to really, truly, break it—my mouse was good and dead. Thanks to miracles in modern crafting, I can now turn my dead mouse into a musical Mouseraca and shake it in joy rather than frustration.

MATERIALS

- Broken trackball mouse
- Phillips-head screwdriver
- Diagonal cutters
- Hot-glue gun
- Dried beans
- Rice
- Cardboard (the same color or painted the same color as the mouse)
- 1" flat paintbrush
- Ruler
- A permanent marker (the same color as the mouse)

Mr. Resistor Man Says:
The first computer mouse—a wooden case with two wheels that allowed for movement along a single axis—was unveiled in 1968 by Douglas Engelbart at the Stanford Research Institute.

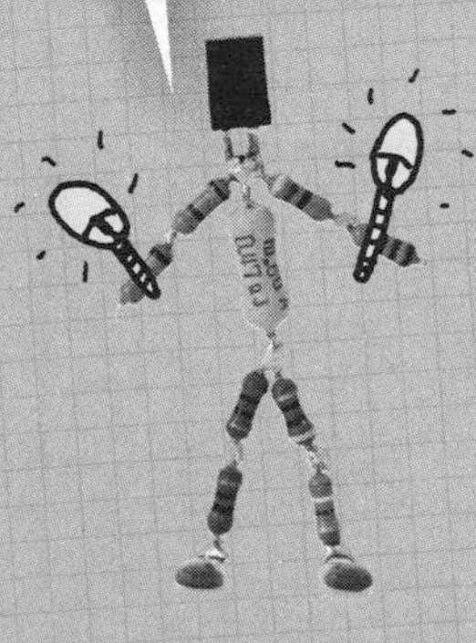

MAKE IT

1. GUT THE MOUSE

Remove the screw on the mouse to open it up. Take out the innards, including the trackball mechanism and gears. Set aside the screws.

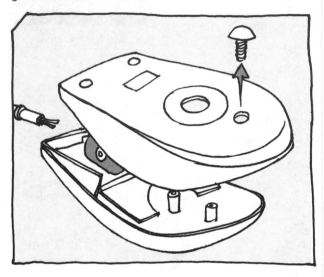

2. CLEAN IT

With your diagonal cutters, cut out as much excess plastic as you can from the inside shell of the mouse. Avoid any plastic that is keeping the mouse buttons in place.

3. REASSEMBLE

Once completely gutted, glue the trackwheel in place with hot glue and put the mouse back together.

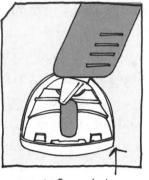

Don't forget to reinsert the screws.

4. STUFF IT

Insert about a dozen dried beans and a few pinches of rice through the trackball hole and then place your thumb over the opening and give it a shake. Add or subtract beans and rice as necessary until you have a sound you are happy with.

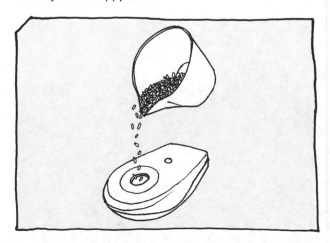

5. COVER THE HOLE

Remove the plastic disc that holds the trackball from the bottom of the mouse and apply hot glue around the opening of the hole. Cover this hole with a piece of cardboard that matches the mouse.

6. GLUE

Cut the bristles off your brush. Hot-glue the brush handle onto the back of the mouse to form a solid handle for your shaker, as pictured.

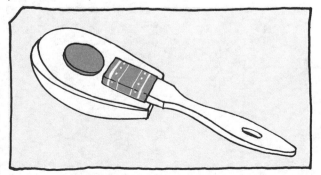

7. CLIP

Find the circuit board that was removed from the mouse and detach the wire.

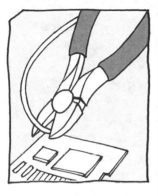

Reuse or recycle the rest of the innards.

8. MORE GLUING

Measure about 2" to 3" down from the plug and glue the length of wire onto the back underside of the mouse where the casing intersects with the base of the handle.

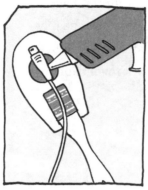

9. WRAP

Tightly wrap the wire around the brush handle until you have reached the end. At the end, trim the wire and glue it in place to the end of the handle. Add some extra squeezes of hot glue to reinforce as necessary. (*Note:* Using a glue stronger than hot glue, like epoxy, will create a more durable mouseraca, but also requires clamping and longer time to dry.)

10. TOUCH IT UP

With your permanent marker, color in any exposed portions of the brush or excess glue.

Shake-shake-shake, shake-shake-shake, shake your... old computer mice!

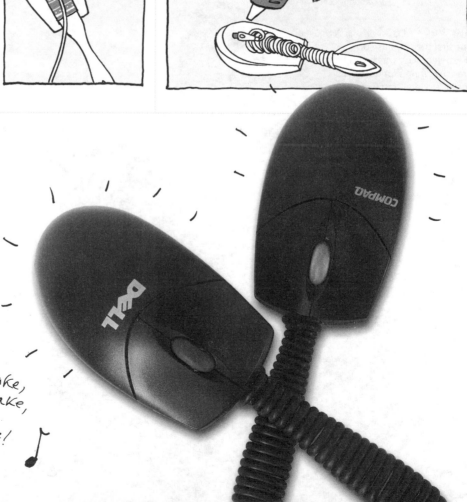

#45 Acoustic ROCK Revolution

LEVEL 3: Expert | **TECH TRASH:** Piezo Transducer

Don't you wish you could combine rocking out with playing the acoustic guitar? Guess what? You can. And you don't have to invest too much money to do it. The days of feedback, distortion, and cranking it up to 11 are no longer purely relegated to the electric guitar. Due to great advances in technology in recent 30 years, you can now cheaply and easily electrically amplify the natural sounds of an acoustic guitar. It is time, brothers and sisters, to join the acoustic rock revolution.

MATERIALS

- Piezo transducer*
- Mini screwdriver
- Razor blade
- Stranded hookup wire
- Ruler
- Wire cutter/stripper
- Soldering iron setup
- Marker
- Acoustic guitar
- Power drill with a 3/8" wood bit
- Two-sided tape
- Solid thick wire or coat hanger
- 1/4" mono jack

Mr. Resistor Man Says:
The first practical use of piezoelectricity was by the French during World War I as sonar for detecting German submarines.

Ms. Resistor Woman Says:
*Find a piezo transducer in an old alarm clock, phone, cell phone, PDA, computer, or most anything that beeps. A broken alarm clock or cheap toy that beeps is always a safe bet. To quickly tell the difference between a piezo and a small speaker, take something magnetic and see if it sticks. If it does, then there's a magnet inside and is a speaker. If not, you've found a piezo.

MAKE IT

1. PREPARE THE PIEZO

Remove the small, flat piezo element (and all attached wires) from the source device while being cautious not to bend it. If the piece you are working with is not perfectly flat, there is a casing attached. Use a mini screwdriver and/or razor blade to gently remove all casings while being extra-careful not to bend or scratch the center of the disc.

2. ATTACH SOME WIRES

Attach about 2' of red and black wire, respectively, to the red (positive) and black (ground) wire coming off the piezo element.

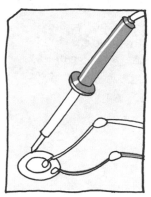

Piezo Particulars
If the wires on the piezo are not red and black, you can tell apart positive and ground by looking at their attachments. Ground is attached to the larger outer ring of the element, whereas positive is attached to the smaller, inner ring.

3. ATTACH THE JACK

Solder the red and black wires to the jack so that black connects to the ground pin and red to the audio pin. *Note:* If you're not sure which pin is which, look where the pins are connected. The black audio wire should connect to the pin that is attached to the bit of metal that encircles the inside of the plug. The red wire should be connected to the tab that is attached to the part of the jack that extends out the back (the part that gets pushed back when you insert the plug).

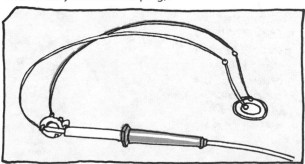

4. MAKE YOUR MARK

With a marker, draw a small dot on the guitar where you want your ¼" jack to be installed. Typically you want to install on the bottom right side of the guitar body (as if you were holding it).

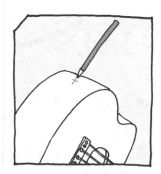

* The acoustic rock revolution will not be televised.

continued >>

5. HOLD YOUR BREATH AND DRILL

Using ⅜" wood drill bit, make a hole where the marking is.

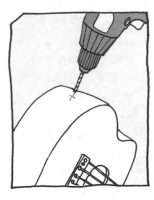

HELPFUL HINT

The trick is to have the bit spin fast while the drill moves in slowly. In other words, don't apply too much pressure on the drill, and take your time pushing it through the guitar to avoid cracking the case. Let the drill bit slowly bore its own way through.

6. PULL SOME STRINGS

You're going to need to stick your arm into the body of the guitar in subsequent steps. If your arm is small enough, you can probably get away with loosening the four highest strings rather than removing them all. You can tell that the strings are being loosened because they will get lower rather than higher as you turn the tuning keys. Otherwise, remove all the strings.

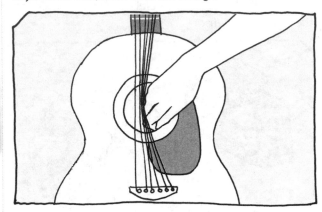

7. ATTACH THE PIEZO

Place a piece of two-sided tape on the flat, wireless side of the piezo. Stick it inside the guitar directly underneath the bridge (the aptly named part of the guitar that holds the strings stationary and provides them with elevation).

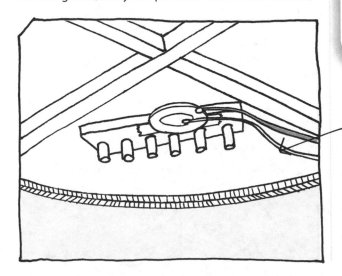

You can tweak the sound the piezo produces by positioning it to hang slightly off the edge of the underside of the bridge.

Why This Works

You place the piezo under the bridge because the bridge is responsible for transferring the vibrations from the strings into the body of the guitar. When the guitar vibrates, the piezo makes an identical electric signal, which can be translated to the amplifier as sound. You stick your piezo to the place that vibrates the most to get the strongest electric signal.

8. A GUIDE FOR DARK PLACES

Take your solid thick wire or coat hanger and fashion it into a straight shaft. This will be used as a guide for getting the ¼" jack out of the inside of the guitar.

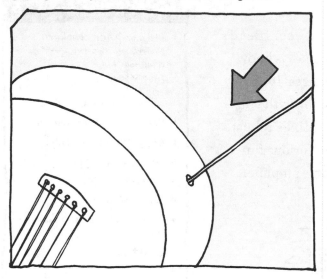

9. OH, NUTS!

Remove the nut from the ¼" jack. You will need this to fasten it in place once the jack is passed through the hole.

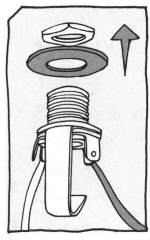

*Inserting the jack may take some time and patience. Stay calm.

10. INSERT THE JACK

Insert your guide through the hole with one hand. With your other hand, slide the jack onto the guide (the thick wire) and use it to bring the jack all the way down through the hole. Once the threading is poking out of the guitar, you can then screw the nut back onto the jack to secure it in place.

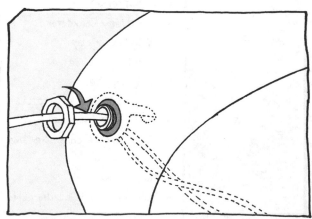

11. ROCK OUT

You've now installed your pickup. The only thing left to do is retune (tighten all those strings), jack in, crank up, and rock out acoustic style.

Now go book some gigs!

ACOUSTIC ROCK REVOLUTION

>> Variations

*   A guitar is not the only thing a piezo can amplify. If you stick it to a drum, you've made a contact mic. If you stick it to a telephone handset, you've made a phone recorder. If you stick it to the outside of a soup can, you've made a tinny-sounding microphone.

#46 Portable Amplifier

LEVEL 4: Geek Squad | **TECH TRASH:** Old Computer Speakers

If you're like me, whether you are in the park, at your friend's house, or lost in the depths of the sewer, you *always* know what time it is because the time for you never changes: It's party time! And no matter where you go, the party is going with you. So, it is important to have a sound system that's mobile, robust, and durable enough to keep up with your fast and furious fun-paced lifestyle. And so I present to you: the portable amplifier.

MATERIALS

▸ **Old computer speakers** (Preferably with a 9V plug—If it reads more, that is okay. You can probably power an amplifier up to 15V with a 9V battery. Avoid less than 9V with a 9V battery, as it may overheat.)
▸ **Phillips-head screwdriver**
▸ **Computer power supply**
▸ **Pencil and paper**
▸ **Diagonal cutters**
▸ **Scrap of wood**
▸ **Power drill with a 3/8" drill bit**
▸ **Craft knife**
▸ **Ruler**
▸ **Craft glue or a lighter** (see Step 5)
▸ **A foot of 1" wide cotton or nylon webbing**
▸ **Four 1/4" eyelets and eyelet fasteners**
▸ **Clamp**
▸ **Four 1/4-20 by 1" bolts**
▸ **Four 1/4-20 nuts**
▸ **9V battery**
▸ **9V battery connector**
▸ **9V battery clip**
▸ **Soldering iron setup**
▸ **12 4" zip ties**
▸ **Epoxy suitable for metals**
▸ **6-32 by 1/2" nut and bolt**
▸ **12" square of cork sheeting**
▸ **Hot-glue gun**
▸ **Music player**
▸ **Male to male 1/8" audio cable**

MAKE IT

1. TAKE APART THE SPEAKER

Select the speaker with the volume knob and open it. Inside there should be a circuit board that is the actual amplifier and also a speaker. Carefully remove both of these from the case without breaking or disconnecting any wires.

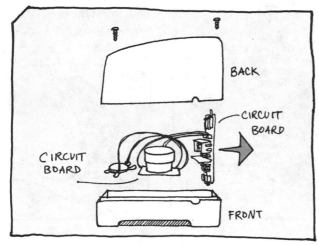

2. AN EMPTY SHELL

Open your standard power supply and remove all of the circuitry except the power source selection switch and power sockets. These are the big switches and sockets attached to the side of the case that give your amp that "electronic" look.

3. ARRANGEMENT

Place the circuit board and speaker inside the case, leaving as much room between them as possible. Align the speaker with the fan grates inside the case, and arrange the circuit board around the speaker to give the speaker some room. Note where the volume and other related knobs will need to be inserted through the side of the case, and where the audio cable will be plugged in. Determine where to drill these holes by measuring the circuit board in relation to the casing and marking the spot at which the two will align.

Note: The bottom of the case will eventually be lined with cork, so the circuit board and volume knobs will be slightly elevated.

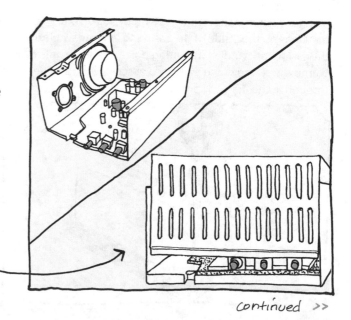

continued >>

4. RIP IT UP

If possible, remove the grates to pass knobs and/or cables through. If the grates can't be removed, clamp the case onto a piece of scrap wood for drilling. Then drill, with a ⅜" bit, through the side of the case at the markings.

5. GET A GRIP

Cut the handle material to the size of a decent hand grip (cut about 6" to 8" to be able to comfortably grab it when mounted to the top of the case). To prevent unraveling, quickly heat each end of the nylon strap with a lighter or, if your strap is made of cotton, cover each end with a small amount of craft glue to make it rigid.

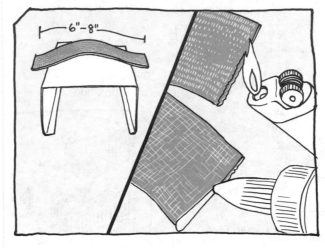

6. MAKE HOLES

On each side of the handle, mark two dots about 1" apart and parallel to the ends of the nylon piece. Then make small holes with your craft knife at each of the small dots. Insert the eyelets through these holes and clamp them.

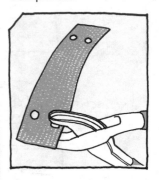

7. DRILL

Using the holes in the handle as a guide, make four markings on the inside of the casing to indicate where in the casing the handle's mounting holes should be drilled. Clamp the casing down to a piece of scrap wood and drill through at the marked spots.

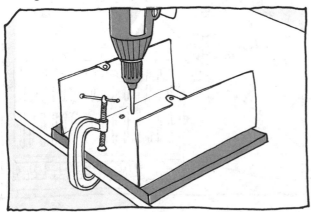

8. BOLT

Attach the strap to the top of the case with four ¼-20 nuts and bolts.

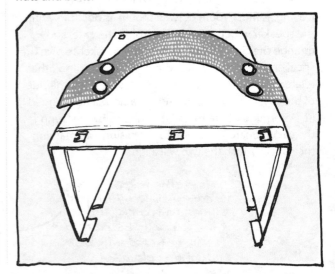

9. WIRE IT UP

Before you can attach the wire that is connected to the battery clip to the circuit board you will need to determine which pins to connect the wires to. There should be two metal pins directly under the 9V power socket on the speaker board. One of these pins will have tracings that are connected to many other pins on the circuit board. This is likely ground. The other is the positive terminal. To test this, plug in a battery and briefly touch each of the exposed wires coming off the battery connector to each pin on the circuit board.

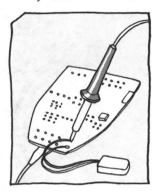

If your speaker turns on, then you have just figured out how the wires need to be connected. If it does not, reverse them. Once you have determined which pins are positive and ground, carefully solder the battery clip wires to the circuit board.

10. MOUNT IT

Securely attach the speaker to the side of the case with zip ties. *Note:* If it's impossible to use zip ties because the holes don't line up, mark and drill mounting holes in the case. Then attach the speaker with nuts and bolts. If the speaker lacks mounting holes to begin with, fasten it to the case using epoxy suitable for metals.

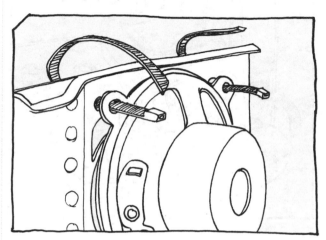

11. INSULATE

The case is conductive, so you are going to want to insulate it. A good way to do this is to cover the bottom with a layer or two of cork (to elevate the circuit board, as needed, to match the elevation of the holes drilled for the volume knobs). Mount the circuit board on top of the cork and loosely attach it to the case using zip ties. Once the case is closed, you should pull these tight to hold the circuit board in place.

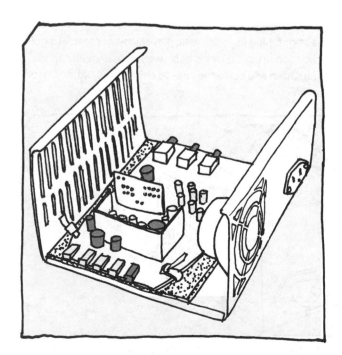

If the cork elevates it too much you can use thin plastic, like the type used for milk cartons. (You can also cover it in electrical tape, but I don't recommend this because over time it may peel and results in shorts.)

continued >>

12. POWER!

Make certain that your amplifier is turned off, unless you plan on using it right now, and plug in your 9V battery.

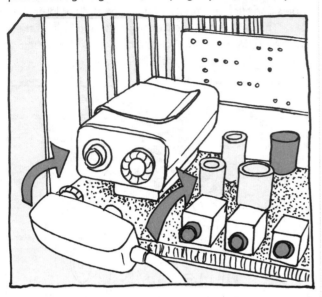

Troubleshooting

If it doesn't turn on at all, first try turning on your speaker. If your speaker is turned on, try a new battery. If that doesn't work, try other pins directly below the 9V power socket.

Amp it up!

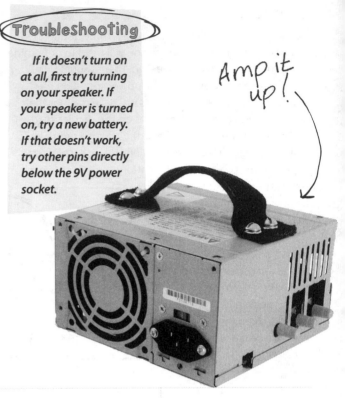

13. CASE CLOSED

Close up the case, pull the zip ties tight, plug in your music player, and get ready to go! *Note:* The reason the case is held closed using zip ties is because you may need to shift the circuit board around when you open and close the case to change the battery. If it is glued in place, this may become difficult.

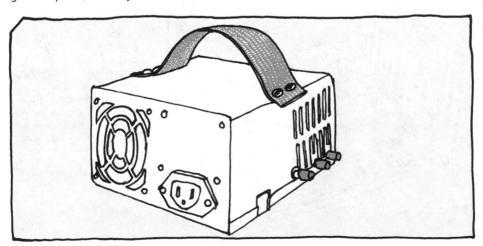

Extra credit!

To use this amplifier with a guitar, you will need to buy or build a preamp circuit and install it in your case. You can also add wires to the circuit board and extend the input jack to the side of the case.

#47 GAMER Guitar

LEVEL 3: Expert | **TECH TRASH:** Broken Nintendo Entertainment System

Your old 8-bit NES gaming console may no longer work, but that doesn't mean that you have to stop playing it. With a couple of modifications, a plank of wood, and some acoustic guitar strings, you can turn your old baby into your new baby: a cigar box style guitar. The best part is that the Gamer Guitar is much healthier to make than a cigar box guitar since no one has to first smoke an entire case of cigars—though you may have to put in some quality time—(on the couch, in front of a TV screen)— playing Super Mario Bros., which is *clearly* much better for you. However it works out, they do say that playing an instrument increases a person's mental faculties, so all you need to do is practice playing the theme song for the *Legend of Zelda* for a couple of thousand hours and it all should balance itself out.

MATERIALS

- A Nintendo Entertainment System
- Phillips-head screwdriver
- Ruler
- Marker
- Power drill with 3/8" and 5/32" drill bits
- Diagonal cutters
- Hacksaw
- Sandpaper (or sanding block, optional)
- 1 x 2 poplar beam, cut 3' long
- 1/2 round pine, cut 1' long
- 3 tuning keys
- 3 metal nuts
- Epoxy
- Pliers
- Hot-glue gun
- Set of classical guitar strings

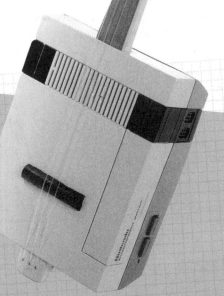

MAKE IT

1. GUT IT

Remove the screws from the bottom of the Nintendo case and open it up. Unplug the controller sockets but leave them installed in the case. Unfasten and remove the rest of the electronics. Set aside the screws as well as the power and reset buttons for later reassembly.

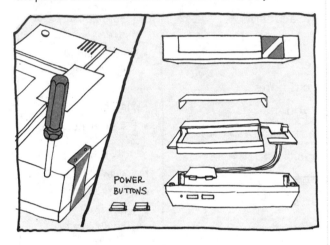

2. MEASURE

On the top of the case, measure and make a centered mark 3⅛" in from one of the short edges. From the same edge, measure and mark ¾" down from the top of the case. Repeat both measurements on the opposite side.

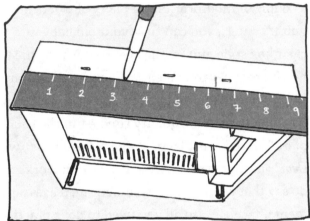

3. CUT HOLES

Drill two holes side by side, ⅜" inside of the measurements you just made. Slowly cut away the plastic within these marks using your diagonal cutters, hacksaw, and sandpaper until you have a rectangle just large enough to pass the poplar beam through. Repeat on the other side so that the two holes line up.

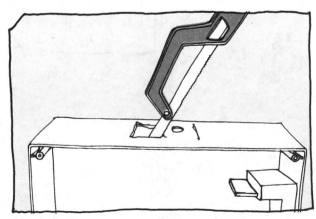

4. GRATED

With diagonal cutters, remove the grates from the inside of the case in the area where the wooden beam will pass over it. Sand the plastic down so that when you pass the wooden beam through the two holes, it can lay flush with the case.

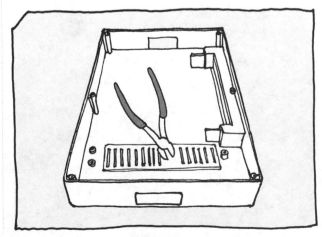

5. A LITTLE OFF THE TOP

Use a saw to "shave" ⅛" of wood off the wider surface 4" from the end, as shown. After sawing in 4" from the end, remove the piece of wood by sawing down from the top ⅛". Sand the cut surface smooth.

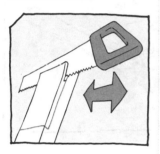

6. DRILL HOLES

Make two staggered marks on the notched surface (from Step 5) at ¼" and ½" in along the left side. Make another staggered mark, ⅜" in along the right. Drill holes at the marks with a drill bit sized for your tuning keys. *Note:* You will likely need to use a ¼" drill bit.

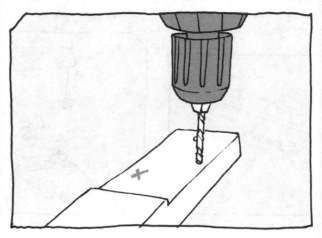

Forget a Les Paul— you're halfway to finishing your N-E-S Paul.

7. DRILL MORE HOLES

Measure ½" in from the opposite side of the board from the notched surface and make a line across the board. Then measure across the line and make a mark at ⅜", ¾", and ⅛". Drill holes at these three marks using a 3⁄32" drill bit.

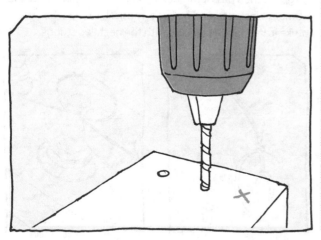

8. FASTEN TUNERS

Insert one of the tuning keys into the beam (the guitar head) from the flat side through to the notched side. Slip on the washer that may have come with the tuning key and then fasten it on with the nut that also came with it. Repeat with the remaining keys. Tighten the nuts with pliers as necessary so they are firmly in place and do not spin.

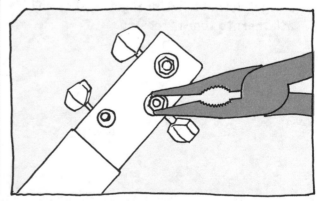

9. PASS IT THROUGH AND GLUE

Flip the top of the NES casing upside down. Also, flip the poplar beam so that the tuning pegs are facing down. Insert the side opposite from the tuning pegs into the hole closest to the grating. Pass it almost completely through, stopping short of the next hole. Apply a generous amount of epoxy to the middle of the case, but avoid getting any on the last two inches of the beam. Before the epoxy starts to set, pass the beam through the second hole, allowing 1½" to extend at the end.

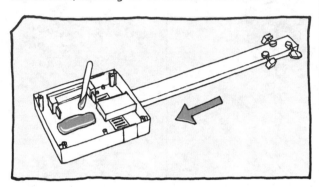

Get a Tune-up!

If you have never tuned a guitar before, the easiest way to tune it is with an electronic guitar tuner that visually shows you when the string is in tune. If you are courageous and want to blow some minds, bring it down to the local guitar shop and ask them to tune it for you.

10. BRIDGE AND NECK

Cut a 3" length of ½ round pine. Use epoxy to glue it in parallel to and 3" from the edge of the case without the grates. Cut a 1½" length of ½ round pine and attach it with epoxy at the edge of the notched section of wood. Optional: Cut notches into the ½ round pine pieces so that when the strings are added, they will be lower and closer to the "fretboard" making them easier to play. This is called lowering the action.

11. STRING IT

Tie one end of the string to a washer. Pass the opposite end of the string through the small hole from the underside and out over the body of the guitar. Next pass the string into the hole in the tuning peg, pull it taut, loop it around the peg, and then pass it through again. Turn the tuner until the string is tensioned and makes a "musical" sound when you strike it. Repeat this process with the next two strings.

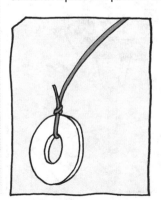

12. CLOSE THE BODY

Hot glue the power and rest button back in. Put the two halves of the casing back together and fasten it shut with the screws you removed earlier.

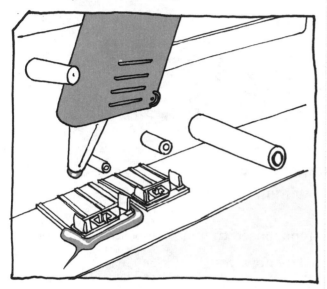

13. CLEAN IT UP

Sand the underside of the guitar's neck (the side opposite the "fretboard") and smooth and round the edges as appropriate so that the guitar is nicer to play and won't leave nasty splinters when you're really shredding. *Optional:* To fancy it up, paint the bridge (the 3" ½ round pine piece) of your guitar black to match the NES. Then lay down some veneer along the top of the fretboard to make it look more finished.

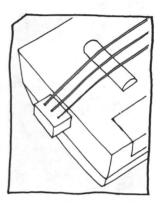

GAMER GUITAR

>> Variations

* Add frets by laying screws across the fretboard and rubber-banding them in place.
* Decorate your guitar with detailing or decals.
* If an acoustic guitar isn't for you, this guitar can easily be amplified using the method illustrated in Acoustic Rock Revolution. Simply stick the piezo anywhere on the inside of the console's casing. Experiment until you get a sound that pleases you.

Extra credit

If you really want to do something cool, you can attach a male ⅛" mono plug to the red and black wires on the end of a NES gamer controller. Then attach the piezo (see Acoustic Rock Revolution, page 184) to amplify the guitar on the inside of the case and wire this into the socket for the controller you left installed in step 1. This way, if you want to amplify the guitar, you can plug in the cord as first player.

You can also add electric guitar effects (fuzz pedal anyone?) inside the body of the guitar and trigger them by pressing the power and reset buttons.

CHAPTER 7

Gadget Goodness

> From the practical to the prank-worthy, here are
  10 projects to unleash the gadget god within.

Y ou've taken apart broken gadgets and repurposed them
to make all kinds of new craft creations. By now, you
should be starting to formulate a rudimentary idea
of how these devices are assembled, what they are assembled
with, and roughly how they function. Perhaps the basic inner
workings of a computer are starting to synthesize in your mind.

Now that these things are beginning to make some sense, it's
time to take a great leap forward and advance to the next level.
In this chapter we will be using our broken gadgets to make new
working gadgets.

> **Mr. Resistor Man Says:** Cartoon detective Inspector Gadget received his bionic transformations after slipping on a banana peel and falling down some stairs.

Postindustrial
Night-Light, p. 228

PROJECTS

As you've already discovered in projects like the Alien Appreciation Key Chain (page 134) and the Resistor Pillow (page 166), dead computers are *full* of functioning parts. Generally speaking, when a computer dies, there are only a handful of key components that stop working. The small periphery components tend to outlive the lifespan of the computer. This is great, because these perfectly functional parts can still be repurposed for their original function, hence enormously extending their life expectancy through the creation of new gadgets. This is the ultimate form of recycling because the components being used retain their initial value for the lifespan of an entirely new gadget.

Better yet, when these gadgets die, you'll have a better understanding as to why and will be more capable of repairing them. Even if your new

Cell Phone Flashlight, p.212

Musical Hold Box, p.224

IR camera, p. 200

DIY Digital Projector, p. 203

magnum

VERNER PANTON

THE NATIONAL GEOGRAPHIC SOCIETY
100 YEARS OF ADVENTURE AND DISCOVERY

gadget is extremely dead and beyond repair, you will know how to take the remaining functioning parts and reuse them to make *even more* gadgets! As for the nonfunctioning components, we already know that they can be repurposed for an array of craft projects (a hair clip, page 102, or corsage, page 109, anyone?).

But by reusing working components and recycling broken ones, you can extend the life of many of the electronic compnents for an exceptionally long time. And so goes the plugged in and battery-powered circle of life: When these new gadgets break, you can disassemble them and make even more gadgets! And when those gadgets break, you can make even more gadgets! And when those break . . . well, I think you get the point.

Now where is your screwdriver? Let's get started.

#48 IR camera

LEVEL 2: Intermediate | **TECH TRASH**: Digital Camera

Over the years, I have collected a number of digital cameras that are not quite broken, but are definitely no longer quite working as they should. And as it turns out, a somewhat-broken camera is the perfect device for dabbling your feet in camera hacking. You already don't expect it to work exactly as it should, so if you make a mistake, there isn't the greatest loss. On the other hand, when you succeed in modifying it, the results are often phenomenal and result in experimental pictures that often far exceed all expectations.

With this in mind, I present to you a beginner's camera hack. With a little bit of alteration, you can convert your camera to take pictures with nonvisible light in the near-infrared spectrum. The results are both fun and spectacular.

MATERIALS

- Digital camera that can still take pictures
- Small screwdriver (for opening the camera)
- Needle-nose pliers
- Scissors
- Toothpick
- Craft glue
- 3" processed, but unexposed, 35mm color film negatives (This is the part at the beginning and end of a processed roll of film that is neither white nor black.)

Mr. Resistor Man Says:
All digital cameras have an IR filter built in to prevent infrared light from reaching the CCD. To make this near-IR camera, you are going to need to remove this filter so that infrared light is allowed to pass through.

MAKE IT

1. OPEN THE CAMERA CASE

Using your screwdriver, carefully open up the plastic camera case and remove as much of it as you can. Don't unplug any wires or cables. Neatly set aside the screws for later reassembly.

2. LOCATE THE CCD

The CCD is the sensor array that takes the pictures. It should be located on a circuit board directly behind the lens and will look like a small box with a shiny glass front. To get to it, carefully separate the lens assembly from the circuit board. This may require removing a few key screws from the assembly.

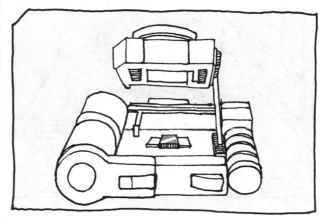

3. LOCATE THE IR FILTER

The IR filter is a small piece of glass that gives off a distinct red and/or blue glare when tilted back and forth under any normal light source. All digital cameras are built differently, but the IR filter is usually located in one of two places and formats: as a small glass square directly on top of the CCD sensor or a small, round glass piece directly behind the lens assembly.

4. REMOVE THE IR FILTER

Use needle-nose pliers or tweezers to carefully pull the IR filter off the camera without scratching the lens or CCD sensor.

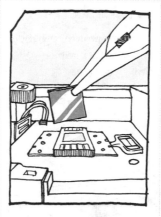

5. MAKE A LIGHT FILTER

Take the 35mm film negative and cut out two identical pieces slightly bigger than the IR filter that you just removed. Try to keep fingerprints and other dirt from getting on the filter, since these may show up in future pictures.

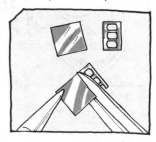

HELPFUL HINT

To view near-IR light, it is important to filter out most light from the visible spectrum. That is what the filter you are making is designed to do. For better results you can try using four to six layers of "Congo Blue" photo filters.

>>>>

6. APPLY GLUE

With a toothpick, spread a minuscule amount of glue on the edges of the frame where you removed the IR filter. Be careful not to get any glue on the CCD or the lens itself, or it will show up in the pictures that you take.

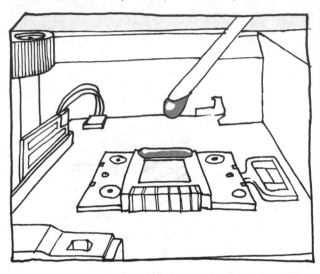

7. GLUE DOWN THE FIRST LIGHT FILTERS

Once the glue is in place, press one of the film filters onto the glue. Again, try to avoid getting fingerprints on the filter or scratching any surfaces. Wait for the glue on the first filter to dry (if the filter shifts while the glue is still drying, you risk getting glue on the CCD sensor or one of the lenses). After you are sure it's dry, drop four tiny dabs of glue in each corner of the first filter and stick down the second filter.

8. REASSEMBLE THE CAMERA

Locate the tiny screws you neatly set aside in Step 1 and reassemble the camera pieces in the order in which you took them apart. Remember to take your time.

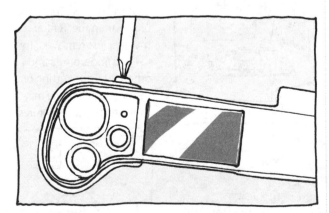

9. TESTING! TESTING!

Go try out your new camera in a variety of environments.

IR CAMERA

>> Variations

★ This modification can be easily done to webcams and digital video cameras. It can even be not-so-easily done to some cell phone cameras.

Try This

You can now take pictures of the light coming from a remote control. In fact, you can even slightly illuminate subjects in what appears to be a pitch-black room using the light from the remote control. To do this, turn off all the lights, point the remote at something close by, press any remote button, and take a picture.

#49 DIY DIGITAL Projector

LEVEL 1: Novice | **TECH TRASH:** LCD Monitor or Laptop

Some people have an extra $1,000 just lying around to buy a digital projector. Unfortunately, most of the people that I know, myself included, do not. And there is nothing worse than being broke and spending long nights on the couch eating cold leftover pizza and watching spaghetti westerns on a 15-inch tube TV. Okay, to be fair, you may like cold pizza, but a screen six feet larger surely couldn't hurt the situation, right? At the very least, you can give Clint Eastwood the justice he deserves, and you won't even need to rob a bank or hold up a stagecoach to make it happen. This DIY projector can be made for $150 or less (depending on how resourceful you are). Worst-case scenario, you may have to eat buttered noodles for an entire week to be able to afford it, but it's a doable short-term sacrifice for an extremely valuable long-term investment in large-scale entertainment.

MATERIALS

- LCD monitor or laptop with a broken backlight
- Mini screwdriver set
- Cardboard box, the same height as the projector bed
- Overhead projector (available on eBay or Craigslist, or at Office Max)
- Small fan
- Four 1/2" spacers
- DVDs!

MAKE IT

1. CASE IT

Remove the screen (LCD assembly) from the plastic laptop casing.

Some LCD screens are difficult or impossible to use because the circuit board is attached by a ribbon cable directly behind the screen. If this turns out to be the case with your laptop broken, fret not; you can still use the others parts of the laptop to make an array of projects.

2. REMOVE THE BACKLIGHT

Once the LCD assembly is free, carefully remove the backlight panel from the assembly. Be careful not to break any electrical connections. You should be left with a brownish glass panel that you can just barely see through.

4. COOL IT

Place a fan next to the projector and position it so that it blows air between the LCD and the bed of the overhead projector.

3. HIGHER GROUND

Use a box or pile of books to elevate the computer or LCD circuitry so that it is level with the bed of the overhead projector and the LCD can lay flat atop the projector bed. Place your ½" spacers under each of the screen's corners, to elevate the screen above the bed of the projector.

Note: The screen should be right side up or your image will be backward!

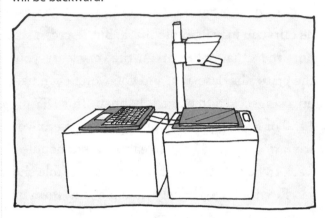

5. READY, SET . . .

Turn on both the projector and the computer. Once booted, insert your DVD.

6. ACTION!

Turn off the lights and focus the overhead projector. Pop some popcorn (kernels are supercheap!) and settle back for an evening of the masterworks of Sergio Leone.

HELPFUL HINT

For a sharper image, try covering up any large light leaks around the bed of the projector. Also, to protect the screen from yellowing over time from the heat of the projector lamp, you can install an IR and a UV filter inside your projector slightly above your lamp (but this will cost you a little extra hard-earned cash).

#50 Laptop Digital Photo Frame

LEVEL 2: Intermediate | **TECH TRASH**: Laptop

Just because your grandparents are elderly doesn't mean you would write them off. The same could be said for old laptops. They may be old, but come the end of the day, they still have much life yet to live, not to mention a lot of photo albums to share! So listen up: Here's an idea that will please both old grandparents and old laptops. Scan all of your family photos onto the family computer and then mount it on the wall as a digital photo frame!

MATERIALS

- Old but working laptop
- Frame, slightly larger than the laptop
- Phillips-head screwdriver
- Mini screwdriver set
- Diagonal cutters
- Gaffer's tape (or blue painter's tape)
- Ruler
- Pencil
- Mat board
- Craft knife
- Double-sided tape
- Power drill
- 3/8" drill bit and a 5/32" drill bit
- Hacksaw
- Screw eyes (screws with loops on the end)
- 19-gauge steel wire
- 1-3 L brackets
- Two 2" wood screws
- Level
- Digital photos

Mr. Resistor Man Says:
Since the 1960s, Moore's Law has held true that the number of transistors that can be fitted onto a microchip doubles every two years. This means that computers become obsolete just as quickly.

MAKE IT

1. LOAD PICTURES

Load all of the pictures that you want to display on the computer into a single folder.

2. START THE SLIDESHOW

Both Windows and Mac computers should have an image slideshow screensaver that comes with the computer. Turn on this screensaver and direct it to the folder with all of your images.

3. CONFIGURE

Change your computer settings so that the screensaver comes on as quickly as possible and the computer never goes into sleep mode. Save your settings and then wait a minute to see if it works. When certain that it does, turn off your computer.

4. OPEN IT

Unscrew and pry off the plastic casing surrounding the computer screen.

5. DISCONNECT

Disconnect the screen assembly from the computer assembly by unscrewing it from the hinges. Then remove the laptop battery from the bottom so that it will be considerably lighter when you hang it on the wall.

6. FLIP THE SCREEN

Fold the screen backward so that it lies flat on the back of the computer. You may need to remove some of the plastic casing to ensure it lies flat.

7. LINE IT UP

Place the screen in the center of the frame. Make "L" marks with gaffer's tape (or blue painter's tape) on the glass to mark the corner of the screen.

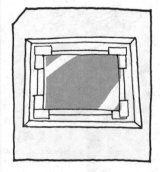

8. MEASURE, MARK, AND CUT

Measure the length and width of the inside of the back of the frame. Using the measurements, neatly draw the outline onto the back of a piece of mat board. Next, draw the measurements for the frame's window centered inside the first rectangle (shown). Then use a craft knife and straightedge to cut along the lines of both rectangles. *Note:* If the inner edge of the cardboard is white, carefully color it in with a black marker.

9. TAPE IT

Check that your mat fits neatly inside the frame and also fits around the screen. Trim it to correct it if necessary. Then line the edge of the screen with double-sided tape and carefully lay the mat over the tape so that it is centered and secure. Set it aside.

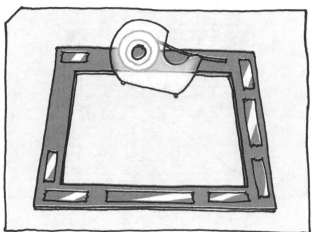

10. VENT

Figure out which side of the frame will be the top. Mark and drill a ventilation hole roughly every 1½" using a ⅜" drill bit.

Drill holes in the top of the frame because 1) it will be harder for people to see the holes when the frame is mounted on the wall and 2) heat rises.

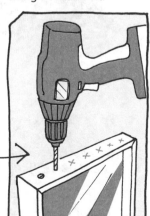

11. PLUG

Identify where the plug will need to come in through the bottom of the frame so that it can plug into the monitor. Mark and drill a ⅜" hole in this location. *Note:* If your plug does not fit through the hole, saw a notch from the back of the frame into the hole. This will allow you to slide the wire in place.

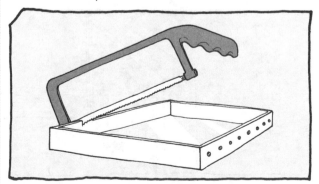

>>>>

12. PUT IT TOGETHER

Insert the laptop into the frame. Plug in the power cord.

13. SAFETY FIRST

Attach small screw eyes near each inside corner of the back of the frame. Diagonally connect the screw eyes with thin steel wire, twisting the wire in place. This will form an X that will hold the laptop in place in the back.

14. MOUNTING

Locate the stud in the wall where you would like to mount your screen and mark where the mounting holes are for the L bracket. Drill $\frac{5}{32}$" pilot holes (or appropriate for the screw size) and secure the L bracket evenly to the wall. (Use a level to measure the alignment.)

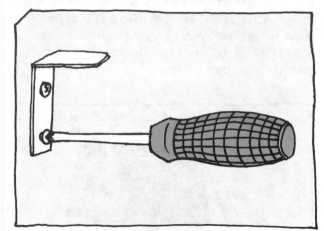

15. TURN IT ON

Turn on the computer and wait until the slideshow starts to play to make certain it is working correctly.

16. THE BIG MOMENT

Place the frame centered upon the L brackets.

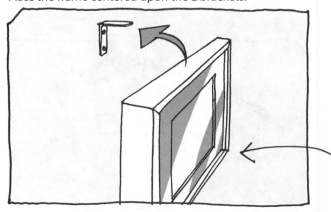

If your frame does not seem to be balancing well on a single L bracket, fasten two more L brackets to the wall on either side of the first one. (These new L brackets will provide balance for the frame as opposed to structural support.)

#51 COMPUTER Phone charger

LEVEL 2: Intermediate | **TECH TRASH:** Phone Charger

Most phone chargers operate at 5V. Coincidentally, all USB ports operate at 5V. If you have an extra phone charger lying around because your insurance plan enabled you to go through five models of the same exact phone, you can easily convert it into a USB charger. And as your phone charges by the mystical power of USB, you will finally be able to stop beating up on yourself about having dropped your phone into the canal in an attempt to have a picture taken as you poked a manatee with your finger. Although, it did serve you right. What did the gentle endangered sea cow ever do to you (aside from racking up $500 in long-distance calls)?

MATERIALS

- Phone charger (make sure it operates at 5V DC by reading the helpful warning label on the plug)
- Craft knife
- Masking tape
- Wire strippers/cutter
- Multimeter
- USB cable
- Soldering iron
- Electrical tape

MAKE IT

1. SPLIT THE WIRE

With a craft knife, split the wire of the phone charger down the middle about 2" so that you have two separate insulated wires running side by side.

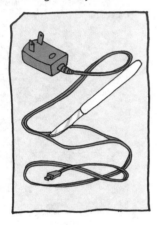

2. MARK THE WIRES

Mark one of the two wires with two small pieces of tape so that later you remember which wire was connected where. You will need to know this when you test for power and ground. If you accidentally reverse the polarity to your phone, you will be sorry! (i.e., your phone will cease to function and your computer may get damaged).

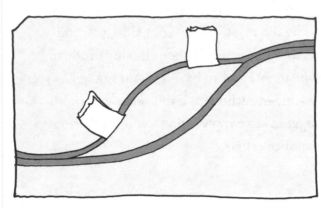

3. CUT THE WIRES

Cut the two wires in half between the two pieces of tape on each respective wire. Strip off about 1" from all four ends of wire. Make sure that they still remain marked.

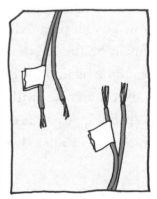

4. TEST THE POLARITY

Plug in your charger and, with your multimeter, test to see which wire is power and which is ground. When you have figured out which wire is which, mark accordingly on the pieces of tape.

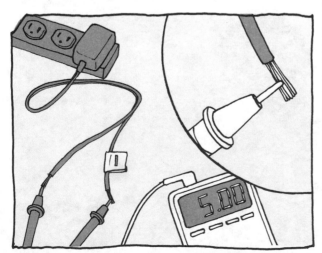

HELPFUL HINT

Remember to make sure the multimeter is set to read the circuit's voltage. When you touch the probes to the wires, it should display a reading of +5V. If you are seeing –5V, then you have the probes backwards and need to reverse them. Remember that red is positive voltage and black is ground (or negative). For a refresher, instructions on how to use a multimeter can be found in the Tools section, page 19.

5. PREPARE YOUR USB CABLE

Cut off the end of the USB cable that doesn't plug into the computer's USB port. Then cut back the cable's insulation jacket. Locate the red and black wires and strip 1" off the end of each. Cut off any extra wires (i.e., yellow and green).

6. KNOT THE WIRE

Tie the red and black wires together in a basic overhand knot to keep your cable from breaking later.

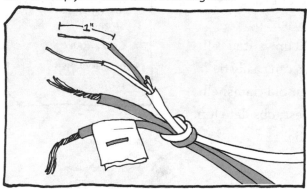

7. MAKE THE CONNECTION

Twist the red wire from the USB cable to the positive voltage side on the phone plug and the black wire to the ground (negative voltage) side. Solder both connections in place. When you are done, you can remove the masking tape markers on the wires.

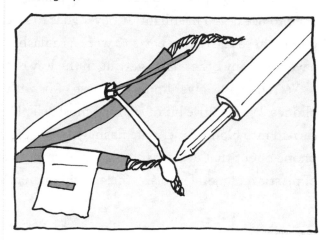

8. INSULATE

Wrap electrical tape around each of the exposed wires individually to keep them from crossing (and killing your phone). Once each wire is wrapped individually, wrap more tape around both wires to hold them together.

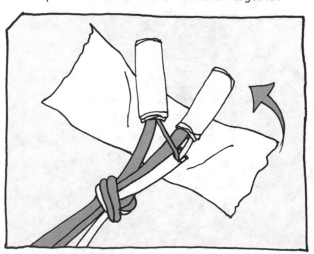

9. CHARGE

Plug in your dead phone to your laptop and charge it up!

Now, the next time the juice Is running low in your phone, introduce it to your laptop!

#52 Cell Phone Flashlight

LEVEL 1: Novice | **TECH TRASH:** Clamshell Phone

E ven though it's not suited to the task, at some point most everybody has pulled out their cell phone, flipped it open, and, tried to use it as a flashlight. It never goes too well. It's not that cell phones have the wrong parts to work as a flashlight, but that they have too many other parts getting in the way of the right parts.

Yet, there is a silver lining for those of us with old unwanted cell phones. With a little bit of surgery, most cell phones can easily be turned into highly functional flashlights. So, dig your old clamshell phone out of that drawer and convert it from a useless, obsolete lump of plastic and metal into an efficient flip-open flashlight.

MATERIALS

- Old clamshell phone (that still turns on)
- Small screwdriver set (or Torx wrench)
- Craft knife
- Hot-glue gun
- Scissors

Mr. Resistor Man Says:
Since the average American goes through a cell phone every 12 to 18 months and only roughly 10 percent are disposed of, it is estimated that more than 700 million cell phones are being stockpiled throughout American homes (begging for reuse).

MAKE IT

1. OPEN

Remove the battery from the phone, then find the small screws holding each half of the phone's casing together. (*Note:* Screws may be hidden under stickers or rubber bits.)

Remove these screws and set them aside for later reassembly.

2. BUTTON PAD

Then take apart the plastic casing and peel the button pad from the phone.

In most phones, the button pad is backlit, therefore, blocking light from shining.

3. SURGERY

Locate the screen assembly and carefully separate the LCD assembly (a dark, semitranslucent sandwich of glass that produces the visual display) from the rest of the screen circuitry. The LCD should still be loosely attached by a thin piece of plastic circuit board or thin wires. Carefully cut away only these attachments.

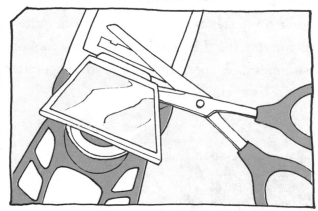

4. FILTERING

There might be a few light filters between the LCD and the backlight. Gather these filters together and carefully glue them back in place with a small drop of hot glue in each corner of the frame.

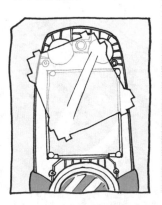

5. POWER BUTTON

Cut away the on/off (often the "End") button from the button pad. Insert this button back into the phone casing.

6. CLOSE

Reassemble the plastic casing and fasten it shut with the screws you removed in Step 1. Reinsert the battery into the phone.

7. POWER

Turn your phone on and flip it closed. When you need light, just flip it open again. Don't forget to keep it charged.

HELPFUL HINT

Conserve energy by keeping your flashlight power completely off when you know you definitely won't be needing it.

#53 USB DESKTOP Fan

LEVEL 2: Intermediate | **TECH TRASH:** Computer's Cooling Fan

Perhaps checking your e-mail gets you all hot and bothered. Or perhaps you just can't afford air-conditioning (or choose not to run it for environmental reasons). Either way, if you are at your computer and need to cool down quickly, a USB desktop fan is for you. It's the perfect way to stay cool on those scorching hot summer days spent indoors at your computer (when you should really be at the beach).

MATERIALS

- Computer's cooling fan
- Safety goggles
- Phillips-head screwdriver
- USB mouse or keyboard
- Diagonal cutters
- Wire cutter/stripper
- Metal coat hanger
- Ruler
- Soldering iron and solder
- Electrical tape

Mr. Resistor Man Says:
A USB desktop fan is basically a computer's cooling fan repurposed as a human cooling fan. This works because the USB port provides 5V of DC power, which is more than enough electricity to power the fan!

MAKE IT

1. OPEN KEYBOARD

Put on your safety goggles. Unscrew the casing on the keyboard or mouse. You may also need to pry it apart at its seam with a screwdriver, since it may also be glued. Don't be afraid to use some force, but be mindful that when the case finally breaks apart, parts might go flying.

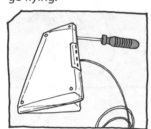

2. LOCATE THE USB CABLE

Find the spot where the USB cable is attached to the circuit board. Usually there are four wires coming from the USB cable, but sometimes there are five. The fifth wire is an extra shielding wire. Use your wire cutters to cut all of the wires free from the circuit board. Try to cut them as close to the circuit board as possible.

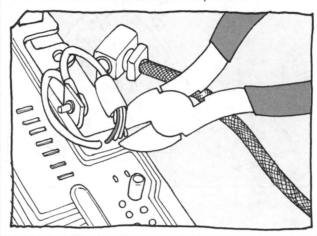

HELPFUL HINT

In some cases, the USB cable is connected to the circuit board with a plastic socket. If this is the case, simply unplug the socket from the board.

3. STRIP THE WIRES

Peel away the insulation on the red and black wires, exposing 1" of wire.

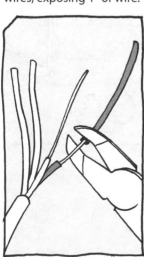

4. DUST THE FAN

Wipe the dust off the computer fan. (Since the cooling fan helped remove dust from your old computer, it's likely covered in grime after years of service.)

5. START MAKING THE BASE

Take the coat hanger and trim it with the cutting wrench 3" away from the hook in both directions.

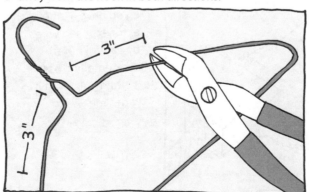

6. INSERT THE WIRES

Insert the two wires from the coat hanger into the top mounting holes on the fan. Push them in as far as they will go.

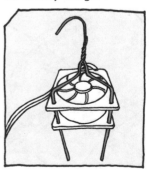

7. START THE BASE

Fold down the hook so that it touches the bottom of the back of the fan (the side air blows away from).

8. CONTINUE THE BASE

Next, bend the hook outward 90 degrees to create a base. Then, fold the entire curved portion of the hook back slightly so that the fan reclines at a nice angle.

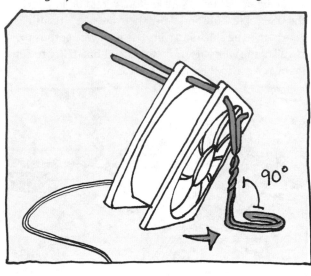

9. TRIM IT CLOSE

Bend back the two metal parts of the hanger that are protruding from the front of the fan. Then use the diagonal cutters to trim the metal part of the hanger so that only ½" remains.

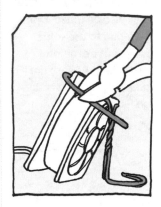

10. KEEP ON KEEPING ON

Plug the red wire from the USB plug into the fan socket terminal that lines up with the red fan wire. Repeat the process with the black wires. If your USB wires are stranded, twist them so they fit into the USB socket. Then apply a small amount of solder to the end of each stranded wire. This will make inserting the wire into the socket easier.

11. WRAP IT UP!

To keep the wires from pulling apart easily, gently wrap your connections with electrical tape.

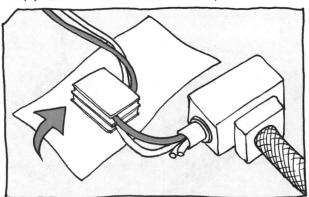

HELPFUL HINT

If you have a lot of extra exposed wire, be careful not to accidentally cross it when you wrap it with tape.

12. BE COOL

Plug your fan in and chill out to the max.

#54 SHINY HAPPY Hard Drive SPEAKER

LEVEL 3: Expert | **TECH TRASH:** Hard Drive

L ike raccoons, people are inexplicably drawn to shiny objects, and a hard drive is about as shiny as they come. It is this shininess that makes it an ideal stereo speaker—face it, kicking out the jams is more about style than it is about fidelity. And just imagine how people will react when they discover that your stereo is pumping through a hard drive—you will have a room full of shiny happy people!

MATERIALS

- Hard drive
- Torx wrench and/ or Phillips-head screwdriver
- Wire cutters/ strippers
- Hookup wire
- Soldering iron and solder
- Amplifier
- Speaker wire

Mr. Resistor Man Says:

A hard drive has the same parts as a standard speaker; namely, a coil atop a large magnet. Even though the hard drive is designed to best function as a data storage device, the placement of the components inside the hard drive easily allow it to function as an audio speaker.

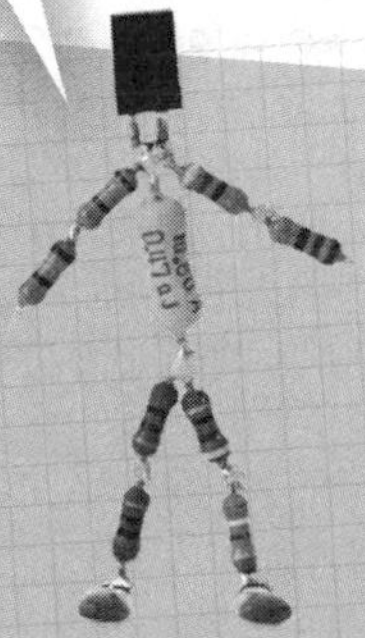

MAKE IT

1. OPEN HARD DRIVE

Using the Torx wrench, remove the screws holding the top cover on. (*Note:* Some screws may be hidden under product labels). After the screws are out, carefully lift off the cover.

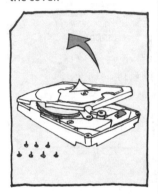

2. FIND COIL WIRES

Locate the tiny coil wires attached to the bottom of the read arm. These are likely located on the bottom of the read mechanism toward the inside of the case. In most hard drives, they are directly connected to a plastic ribbon cable with thin conductive tracings (tracings are those little lines you typically see covering the surface of a circuit board).

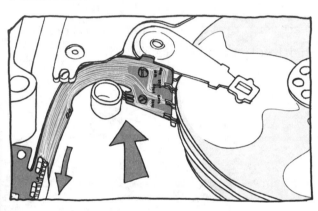

3. TRACE

Once you have found the coil wires, follow the attached conductive tracings to the solder pads on the other side of the hard drive.

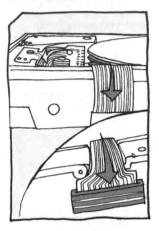

HELPFUL HINT

The terminals will most likely be side by side. Conductive tracings cannot cross each other or be intersected by other tracings. Although the tracings on ribbon cables may seem overwhelming to look at, they are actually easy to keep track of.

4. TEST YOUR TERMINALS

When you think you have identified the terminals that connect to the coil wires, try testing them: Unplug the wires from the back of your stereo speaker and, without crossing them, touch a wire to each terminal. You should hear your hard drive playing music. *Hint:* Twisting the stranded speaker wire and applying a small amount of solder will make it easier to work with.

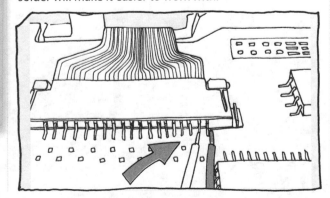

Troubleshooting

If you don't hear music, first check that the stereo is on and playing music and the volume is at a reasonable level. If you have tried this and it still doesn't work, try redoing the previous step.

5. CONNECT THE LEAD WIRES

Cut two 6" lengths of wire and strip them at both ends. Solder one of these wires to each of the two terminals. Make sure that the wires aren't crossing. If the wires get crossed at the solder points, then it won't work.

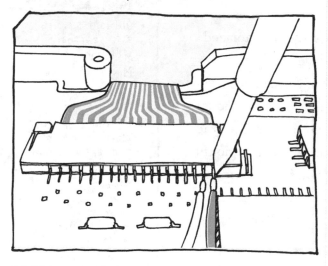

6. HOOK UP YOUR AMP

Twist one of the lead wires coming off the hard drive to one of the speaker wires. Repeat with the other two wires. For a more permanent connection, solder the lead wires and speaker wires together.

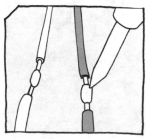

7. PLUG IN

Plug the other end of the speaker wire (the cable) into the amplifier.

SHINY HAPPY HARD DRIVE SPEAKER

>> Variation

* For extra credit: Did you know that you could also play music through the stepper motor that spins the hard drive platters? Poke your stereo wires at the four large solder terminals on the motor and see which pair of wires gives you the best sound.

Once you have figured out which two terminals play music the best, for a larger audio experience, carefully solder the motor in, parallel with the coil wires.

HELPFUL HINT

Apply a small drop of solder to each wire first, then you can easily connect it to the terminal by touching the wire to the terminal and rapidly heating the wire to make a quick connection.

Troubleshooting

"I crossed the wires! Now what?" Ideally, you've connected wires to the back of the hard drive (there are no moving parts that can rip your wires out). With that said, if you have crossed some wires, you can unplug the ribbon cable from the circuit board and carefully attach your lead wires to the terminals on that.

If you cross the coil pins on the ribbon cable, simply cut the ribbon cable in half and connect your audio wires directly to the coil wires on the read arm. This should be a last resort, because if the coil wires break at this juncture, there is little hope in fixing them. Be sure to use a thin stranded wire for your leads and leave plenty of slack to prevent the coil wires from breaking when the read arm moves to the music later on.

#55 BRIGHT IDEA Light Pad

LEVEL 3: Expert | **TECH TRASH:** Laptop

Traditionally, light tables are not light. They are big, fat, and ugly wooden contraptions filled with lightbulbs. They're not easily transported, and they take up a lot of space. Get ready to be amazed: I bring you a lightweight, notebook-thin, low-voltage, sleek, and highly portable light table. Actually, I don't suppose you can even call this a table anymore. This is more of a light pad, and this light pad is awesome. (Almost as awesome as a light saber.) You can use it to trace designs, make animations, view slides, display transparencies, mount the X-ray of your broken wrist to the wall, and, if you're not accident-prone or artistically inclined in any manner, you can still use it as a large night-light. A boring old light table can't easily do all of that. The times they are a-changin'.

MATERIALS

- Old laptop (with a screen that still lights up)
- Mini screwdriver set
- Hot-glue gun
- 1/8" sheet of clear plastic the size of the laptop screen
- 5V cell phone plug*
  *You can tell if a cell phone plug is 5V by reading the power rating on its label. If it is 5V, it should say that it outputs 5V DC.
- Wire strippers/cutters
- Soldering iron set
- Zip tie
- Multimeter

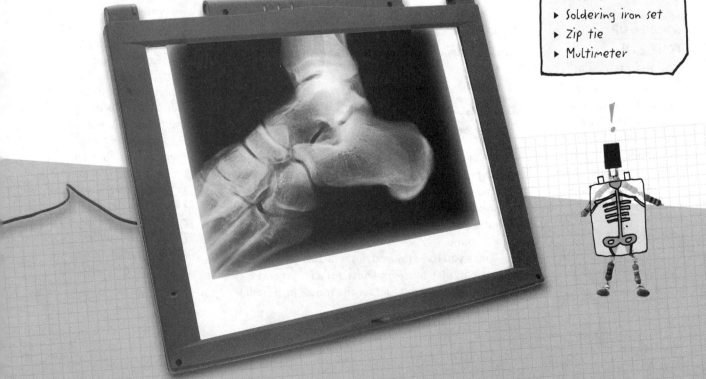

MAKE IT

1. SEEK AND UNSCREW

The screen is being held together by a series of screws. The screws may be being hidden by rubber stoppers or thin plastic stickers. Pick them off and unscrew the screws. Set them aside for later reassembly.

2. OH, SNAP

Carefully snap off the front plastic cover that borders the screen. You may need to use a little bit of force to do this, but be careful not to snap the cover in half.

3. HALVE IT

Now you should have access to the brackets that secure the entire laptop screen assembly to the rest of the laptop. Carefully detach the entire assembly, then detach the screen from the screen assembly.

4. DISASSEMBLE

Carefully disassemble the screen so that you can remove the LCD element.

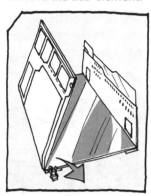

5. REASSEMBLE

Then reassemble the screen without the LCD element.

6. THE PLASTIC SCREEN

Apply a thin line of glue all around the edges of the screen with the hot-glue gun. Carefully secure your ⅛" sheet of clear plastic in the screen's frame before the glue dries.

7. PREPARE THE PLUG

Cut the end off the power plug. Peel back the insulation until the two power wires are exposed. Strip 1" off the ends of these wires.

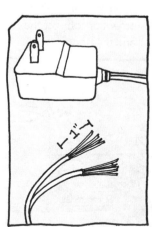

8. LOCATE THE CIRCUIT

There should be a circuit board with two thick wires that connect to the fluorescent tube that illuminates the screen. If you are unsure, it is the board with a large, box-looking transformer near a socket connected to two thick wires (it will most likely have a warning about high voltages printed on it). This is the circuit board that you need to attach the cell phone power source to.

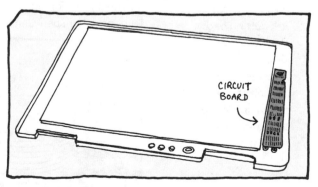

CIRCUIT BOARD

>> >>

9. GROUND

First locate ground on the circuit board. You can usually identify ground by finding a large conductive surface on the underside of the board. This surface should look like there are a lot of component pins soldered to it and be traceable throughout the length of the board. It can be accessed on the top side of the board by figuring out which pin on the bottom of the board connects through to the top.

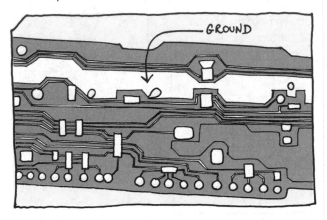

GROUND

10. GET POSITIVE

Once you have located ground, find the place to connect the positive voltage on the coil. The coil should be located in the center of the board before the transformer. In other words, the low-voltage from the cell phone transformer needs to pass through the coil before entering the transformer that converts it to high voltages and powers the tube. Therefore, the place in which you need to connect the positive voltage cannot be physically located on the circuit board anywhere between the coil and the high-voltage wires to the monitor. It has to be before the coil.

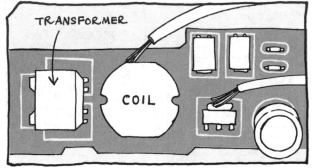

TRANSFORMER

COIL

SAFETY FIRST

>> **Do not touch any of the exposed electrical connections on the board—you risk getting badly shocked. You should also never hold the wires from the cell phone power supply near the conductive metal part.**

11. LET THERE BE LIGHT

Once you think you have found both ground and positive terminals, plug in your cell phone power adapter and touch the ground wire to the suspected ground terminal and the positive wire to the suspected coil. It should light up. If it doesn't work, try reversing them or touching them to some other terminal. Just don't hold the wires in the incorrect place for more than a moment or connect the wires anywhere between the coil and the transformer.

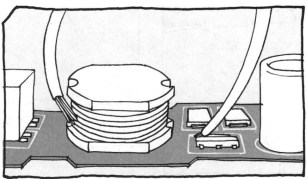

HELPFUL HINT

If you are unsure which wire is positive and which is negative, you can plug in the transformer and test the wires with your multimeter. For instructions on using the multimeter, check the Tools section on page 19.

12. SOLDER

Once you have identified the spot to which you need to connect the wires, solder them in place. Then make sure they work by plugging in the transformer.

13. SECURE

With your hot-glue gun, secure the wires neatly in place so you can reattach the circuit back into the case later on. (In other words, don't cover the screw mounting holes or glob on lots and lots of glue.)

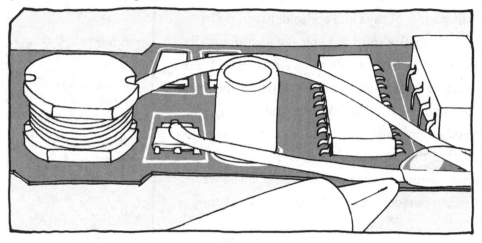

14. ZIP TIE

Make a loop with the wires and pull tight with a zip tie. Cut off the excess plastic strip of the tie.

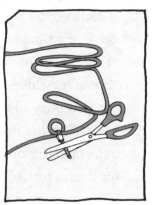

15. TIGHTEN IT UP

Fasten the circuit board back in place using the screws that you put aside in Step 1. Hook the wire loop (with the zip tie) around a piece of plastic or a screw that will serve as an anchor and keep the cord from being ripped out.

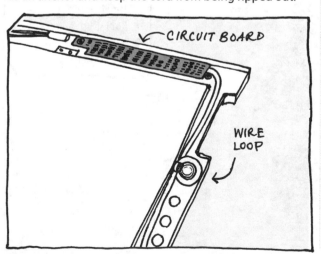

CIRCUIT BOARD

WIRE LOOP

16. SNAP

Snap the plastic shut and fasten it closed with the screws that you removed in Step 1. Plug it in!

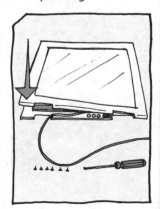

⚡ SAFETY FIRST

>> **Do not touch the circuit board when it's plugged in because of high voltages!**

BRIGHT IDEA LIGHT PAD

>> Variation

* With some minor modifications, you can have this light pad run off batteries!

#56 Musical Hold Box

LEVEL 4: Geek Squad | **TECH TRASH:** External Telephone Modem

D o you ever get the feeling that large bureaucracies hate you? I
do. Every time I call one on the phone to try and accomplish
something, I end up spending most of my time on hold, listening to
horrible, static-ridden music while the soothing voice of a woman tells
me just how busy all of the operators are at this time. And as I sit there
listening to my life drift away in inefficiency, I dream of being able to
place the operator on hold much like they had done to me.

This dream turned into obsession, and, like most good obsessions,
it has now become a reality. Behold, a device that allows you to stick
bill collectors, nagging family members, and people who dial the wrong
number on hold with your own personal musical selection for as long
as you feel they deserve.

MATERIALS

- External telephone modem
- Any parts that you are unable to find in Step 2
- Phillips-head screwdriver
- Desoldering braid
- Telephone cable
- Hookup wire
- Diagonal cutters
- ⅛" mono audio jack
- ⅛" stereo cable
- Soldering setup
- Wire cutter/ strippers
- 220-ohm resistor
- 330-ohm resistor
- DPDT switch
- C-clamp
- Scrap wood
- Hot-glue gun
- Power drill with a ¼" drill bit

MAKE IT

1. OPEN THE MODEM

Use a screwdriver to loosen and open the case of the telephone modem. Remove more screws and take out the circuit board.

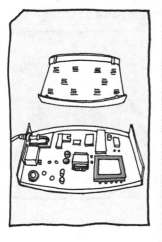

2. LOCATE THE FOLLOWING PARTS

Inside the modem, locate the following parts:

- A 1:1 600-ohm audio isolation transformer
- A 47μF electrolytic capacitor
- Two telephone jacks
- An LED

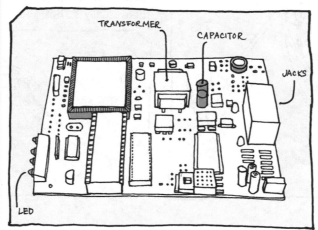

TRANSFORMER

CAPACITOR

JACKS

LED

3. DESOLDER

Flip the circuit board over and, working from the underside, use the desoldering braid (see page 39) to remove the four parts that you located in Step 2 from the circuit board.

4. PREP THE JACK

Flip the two telephone jacks (the double jack) over. Each jack should have four pins sticking out of it. For each jack, take your diagonal cutters and clip off the two outer pins. Bend the two pins that are left slightly away from one another.

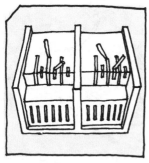

5. JACK TO SWITCH

Face the open jack terminals toward yourself. On the right terminal of the right jack, solder a black wire. On the left terminal of the right jack, solder a red wire. Connect the red wire to one of the middle pins on the switch; the black wire to the other.

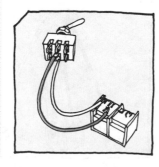

6. AGAIN

On the right-terminal of the left jack, solder a black wire. On the left terminal of the left jack, solder a red wire. Connect the red wire to one of the pins in line with one of the other red wires. Connect the black wire on the pin next to the red wire you just connected.

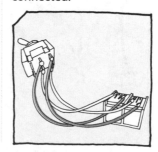

7. CIRCUIT BUILDING

Take both resistors, the positive leg of the capacitor, and a red hookup wire and twist and then solder them together. *Note:* The negative leg of the capacitor is marked with a minus sign. The leg that isn't marked is positive.

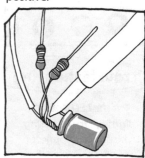

>> >>

8. SWITCH IT UP

Attach the other end of the red wire from Step 7 to the switch. There should now be three red wires, all in a row, attached to the switch.

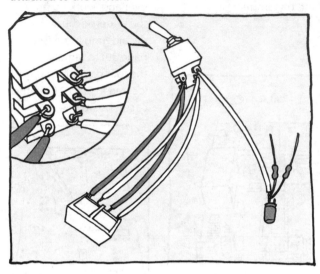

9. TRANSFORMER

Solder a red wire to the negative leg of the capacitor (marked by a minus sign). Attach the other end of the wire to one of the pins on the transformer. (The transformer should only have four pins. If there are five, ignore the pin that appears to be in the center of one side. There should be a coil on each side of the transformer with pins attached on each end of the coil. You can tell the sides apart because there usually is a boxlike structure dividing them.)

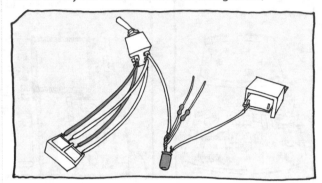

10. TESTING

Test the LED by briefly applying 3V to the two leads. Mark the pin connected to the negative wire.

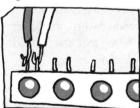

Troubleshooting

If the LED doesn't light up, try reversing the power supply. If it still doesn't work, replace your LED.

11. ON THE PLUS SIDE

Solder a red wire to the end of the 220-ohm resistor. Connect the other end of the wire to the positive leg of the LED.

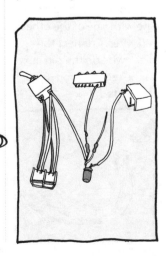

12. EVERYTHING WENT BLACK

Carefully solder a black wire to the negative leg of the LED. Also, solder a different black wire to the last free terminal on the switch. Then solder yet another black wire to the one remaining free pin that is on the same side of the transformer where the red wire is attached.

(If there are two remaining free pins on the side of the transformer that the red wire is attached to, don't select the middle pin.)

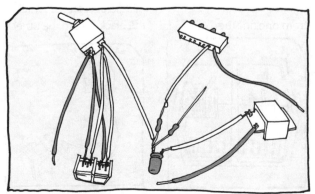

13. GROUND CONTROL

Twist the ends of three black wires together with the free leg of the 330-ohm resistor and solder them to make a permanent connection.

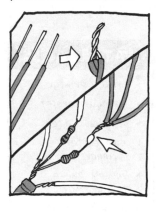

14. THE OTHER SIDE

Attach red and black wire to the other side of the transformer so that it lines up to the red and black wire already on the opposite side.

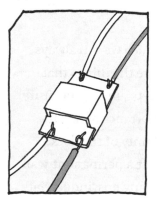

15. AUDIO JACK

Attach red and black wires to the audio jack so that the black wire is electrically attached to the ground tab inside the jack and the red is attached to the other free tab for the audio channel.

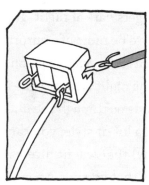

16. DRILL

Check to see if there are existing holes in the back panel of the drive. Otherwise, mark and drill two ¼" holes in the back panel at least 1" apart to mount the switch and the audio jack.

17. PANEL MOUNT

Remove the nuts from the threading on the switch and audio jack. Insert these two components through the holes in the panel and then securely fasten them in place with the respective nuts. Once those are secure, glue the telephone jack and LED properly into place.

18. CLEAN UP

Glue down the circuit inside the case so that it can't move around and the wires won't cross.

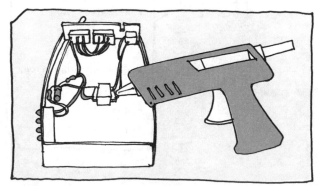

19. PLUG IT IN

Reassemble the modem casing, then plug the phone line coming from the wall into the left phone jack. Plug the one coming from the phone into the right phone jack. Attach your music player to the audio jack.

20. EXACT YOUR REVENGE

Next time you get an unwanted call, start your music player and flip the switch. Just don't forget to turn it off when you are done with it, or your phone line will be left off the hook!

MUSICAL HOLD BOX

>> Variation

* For a more realistic feel, add sound bytes about expected wait times and online billing systems to the mix.

#57 Postindustrial NIGHT-LIGHT

LEVEL 4: Geek Squad | **TECH TRASH**: Broken Printer

I t is no surprise that it gets dark at night. That much we can always count on. What tends to be more of a surprise are the things that lurk in the dark—misplaced shoes, creeping house pets, unruly furniture. These unseen dangers of the night illustrate the importance of having a little safety beacon to guide us through the uncertainty of the dark.

With a little effort and a lot of style, you can craft a permanent wall fixture (or tabletop display) that will fit nicely into any postindustrial loft space. The best part is that you don't have to worry about remembering to plug it in when it surprisingly turns dark at night because this night-light *can turn itself on* when needed. Just install it, and fear the dark no longer!

MATERIALS

- Broken printer
- Screwdriver set
- Diagonal cutters
- Wire cutters/strippers
- 5V cell phone transformer
- Soldering setup
- Electrical tape
- Five white LEDs
- MOSFET (metal-oxide-semiconductor field-effect transistor)
- 1K resistor
- Multipurpose grid-style PC board
- Five 100-ohm resistors
- Solid hookup wire
- Photocell
- Hot-glue gun
- Cork
- Two-sided tape
- Colored tissue paper
- Ruler
- Hammer
- 1¼" wire brads

⚡ SAFETY FIRST

>> **Caution: This project uses high voltages. If you are uncomfortable or unfamiliar working with high voltages, you should not attempt this.**

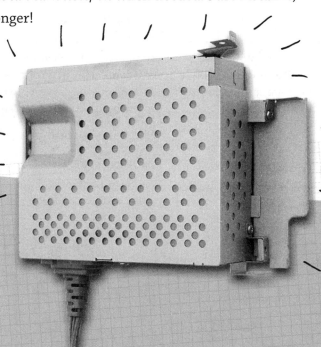

MAKE IT

1. CONTAINER

Open the printer and remove the power supply box from inside.

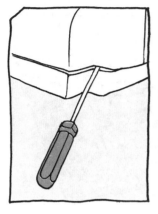

2. EMPTY THE CASE

Open the power supply and carefully remove the circuit board. If necessary, cut the power cable from the board (leaving the cable as long as possible).

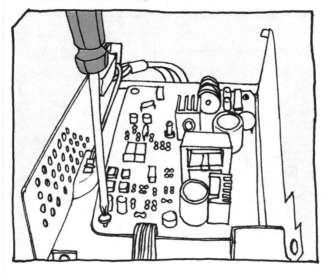

SAFETY FIRST

>> Be extra-careful, because the circuit board can contain charged capacitors that are large enough to kill you. Please reference the Safety section regarding safely discharging high-voltage capacitors.

3. CRACK THE CASE

Break open your transformer and locate the two wires connected to the prongs of the plug. Cut these wires from the prongs and strip the ends to expose the stranded cable.

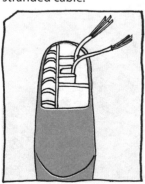

4. HIGH-VOLTAGE

Strip the ends from the two wires coming out of the power cable and individually twist them with the transformer wires (one single wire from one should be connected to one single wire from the other). Solder these connections and carefully insulate each connection individually with lots of electrical tape (at least 6").

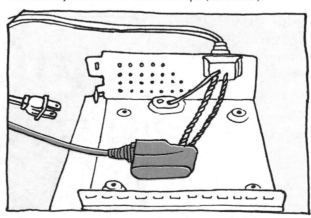

SAFETY FIRST

>> Do not let a single strand of wire escape the insulation (the electrical tape), because that strand could provide the electricity with enough of a path to badly electrocute you.

5. CIRCUIT BOARD TIME

Insert an LED into the circuit board so that the short leg plugs into one of the holes through the long copper strip in the center of the board. The long leg from the LED attaches to the first pin in one of the copper strips with three holes in it that is rotated 90 degrees from the long copper strip. Continue until you have installed all five of your LEDs.

6. RESISTORS

Match up a 100-ohm resistor to one of the five three-holed copper strips that has an LED inserted through it. Connect the other end of the resistor to the other long copper strip in the center of the board, to which the LEDs are not connected. Repeat this step for all five of the LEDs.

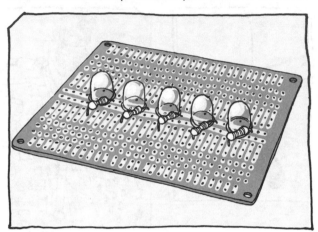

HELPFUL HINT

If you are using completely identical LEDs, you can get away with connecting just one resistor in this manner. However, I don't recommend this shortcut, because if the one LED connected to the resistor stops working, the circuit will no longer have any resistance and overheat.

7. SOLDERING

Solder the component leads to the copper when you are done inserting them. Be careful not to bridge any of the copper strips with solder (if you do, you may short the circuit by crossing the two connections). Trim the component leads with diagonal cutters as close to the circuit board as possible to prevent short-circuits.

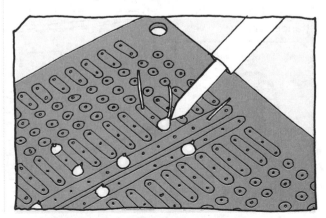

8. MOSFET

Insert the MOSFET into the board so that the three pins go to three different copper strips (all with three holes in them). The pin sticking down on the left side of the MOSFET (when looking at the boxy part) is pin 1. Using a 1K resistor, connect the copper strip that pin 1 is inserted into with the copper strip that all of the 100-ohm resistors are inserted to. The center pin is pin 2. Using a small 1" piece of wire, connect the copper strip that pin 2 is connected to with the copper strip that all of the LEDs are connected to. When you are done, solder all of these parts into place. Make certain nothing else is connected to these copper strips.

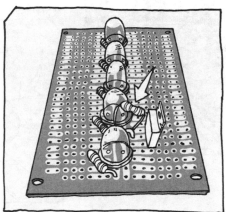

9. PREP

Prepare the photocell by twisting two long black wires to the legs and then soldering each connection.

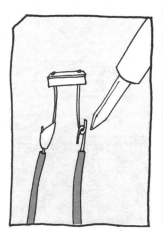

10. PHOTOCELL

Solder one of the wires connected to the photocell to the copper strip that pin 1 of the MOSFET (and the 1k resistor) is connected to. Solder the other wire to the copper strip that is shared by pin 3 (this is the pin with nothing connected to it yet).

Troubleshooting

If your night-light doesn't work, check to make sure that:

1 *You have not crossed the copper terminals with solder on the board and created a short circuit.*
2 *All the wires are connected as instructed.*
3 *The LEDs were inserted with the correct orientation.*
4 *The board is receiving power from the cell phone transformer. Test the low-voltage DC power cables on the LED circuit board with your multimeter.*
5 *You have no solder connections that look like dull, round beads (these are bad connections). They should look like smooth, shiny surfaces.*

>>>>

11. POWER CABLES

Solder a 4" red wire to the long copper strip that all of the resistors are connected to. Solder a 4" black wire connecting the copper strip that is shared by pin 3 of the MOSFET and one leg of the photocell.

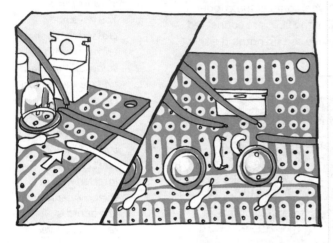

12. LIVE WIRE

Locate the wire that comes off the transformer and plugs into the cell phone. Strip back the jacket 1" to 2" to reveal two colored power cables. One of these wires should be red. Strip a small amount of insulation off the end of this wire and solder it to the red wire sticking off the circuit board. Solder the other colored wire (usually black or white) to the black wire coming off of the circuit board. Insulate these connections separately with a strip of electrical tape.

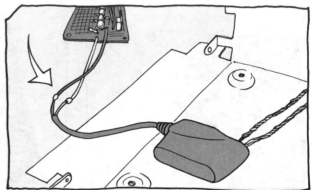

13. CORK

Cover the bottom of the case with cork and hot-glue it in place. (This is to insulate the case and prevent the board from short-circuiting.) Leave any screw holes uncovered so that it can be mounted to the wall.

14. GLUE

Center the circuit board in the case and hot-glue it to the cork. Then arrange the hacked cell phone transformer to the cork, in the case, but off to the side where it will be out of the way.

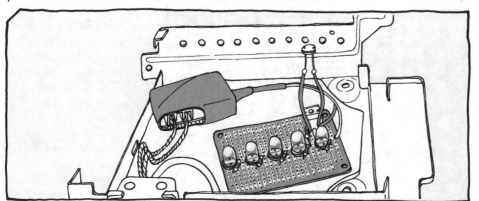

15. TEST

Plug it in. The LEDs should turn on when placed under a bright light and turn off when your thumb covers the photocell.

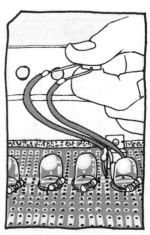

16. LOCATION IS EVERYTHING

The gridlike surface of the photocell is the part that senses light. Hot-glue the photocell to the inside of the case so that the surface is able to sense light. Note the two pins on the surface of the photocell. Keep these two pins spaced slightly apart from the case. (If they are both touching your metal case, it will short-circuit and your night-light won't work properly.) Attach the photo cell where it will be able to sense ambient room lighting during the day (depending on whether you plan to mount it to a wall or plop it on a table).

17. DIFFUSION

Use two-sided tape to lay tissue paper flatly over the holes in the casing. Keep adding layers of tissue paper until you are pleased with how they diffuse the light. Then reassemble the case and fasten it shut again.

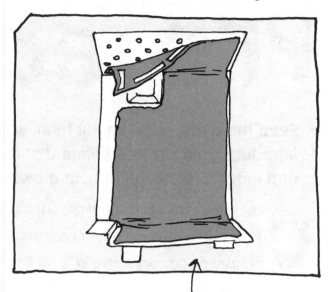

To add an extra bit of diffusion to the front, throw on a trimmed layer of white printer paper.

18. FASTEN

Measure the space between the mounting holes in your casing and make corresponding marks, parallel to the floor, on the wall. Hammer two wire brads at the marks so they stick out a ½" from the wall.

19. PLUG

Mount the casing on the brads and plug in your night-light, and no longer fear the end tables that lurk in the night.

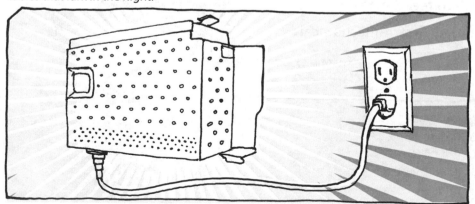

CHAPTER 8

PLAYFUL Pet
Projects

> **Feed the birds, raise an ant farm, wrangle the lizards, introduce your cat to a different sort of dead mouse, and more: 5 projects for your growing menagerie.**

Mr. Resistor
Man Says:
The "Harvard Mouse
patent" (U.S. Patent
No. 4,736,866) was the
first patent granted for
a new animal life-form.

Computers may come and go, but a loyal pet will love you forever. Now, think about how many times you have shunned your pet's love in favor of the cool embrace of that blue-glowing computer screen. How do you think that makes your unquestionably loyal little friends feel? Not very good, I can tell you.

Our pets do not enjoy our computers or any of the assorted electronic devices we have scattered about our homes (except, perhaps, the rare cat who enjoys snuggling up to feel the heat of your old monitor). These strange electro-mechanical

CAT Tube, p. 241

PROJECTS

beings make odd noises, can't be used as toys, and compete (often successfully) for our love and attention. That said, if there is anyone ecstatic when our expensive electronic gadgets kick the bucket, chances are it is our loyal pets.

While it might take your loving pet a little while to snuggle up to the idea, why not use that strange contraption that kept you away from them for so long as a bridge forward to start mending your wounded relationship?

Imagine your cat's astonishment when you cart away that nasty old computer and later present him with fancy new digs made from the old CRT display. Or, perhaps, you pamper him with a tasty snack served inside the carcass of a dead computer mouse. Or maybe you befriend the neighborhood birds by hanging a bird feeder outside your window. Just think about how happy the animals in your life will be with such delightful surprises! And as we all know, a happy pet translates back into a happy human.

So let us convert all those dead computers into something our pets will truly enjoy. Everyone will be the better for it and as a matter of course, it would be irresponsible to do anything less for our special friends. If our pets have taught us anything throughout the years, it is that a little selflessness goes a long way.

iMac Terrarium, p. 248

cat (Grass) and Mouse, p. 239

Flat-Screen Ant Farm, p.243

Stripped-Down Bird Feeder, p. 236

#58 STRIPPED-DOWN Bird Feeder

LEVEL 1: Novice | **TECH TRASH:** Power Strip

This project is thanks to my bird-loving (and slightly birdbrained!) younger brother. It was his idea to turn a power strip into a bird feeder so the birds in his yard could charge all of their "little birdie gadgets—you know, cell phones, MP3 players, that sort of stuff." Never mind that most cell phones and MP3 players are of equal or greater size to the birds that would perch on such a feeder, and the power strip has actually been stripped of all of its electronic capacities . . . I really liked the idea of the power strip as a gathering place of sorts. Rather than charging up electronics, the birds are fueling up for their daily aeronautic adventures!

MATERIALS

- Power strip
- 2 to 3 unwanted 3-pronged power cords
- Diagonal cutters
- Phillips-head screwdriver
- Epoxy
- Pliers
- A funnel
- Zip tie
- Birdseed

Mr. Resistor Man Says:
Some appliances consume "phantom" power, even when turned off. You can stop this from happening (and lower your energy bill) by plugging them into a power strip with an "off" switch.

MAKE IT

1. CUT THE CORDS

Cut the plugs off the power cords, leaving 2" to 3" of cord on the plug. These plugs will be the perches on which the birds will sit. Set them aside for future use.

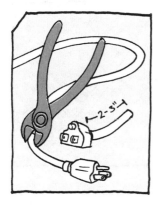

2. OPEN IT

Remove the screws and open up the power strip. Set aside the screws someplace safe for reassembly.

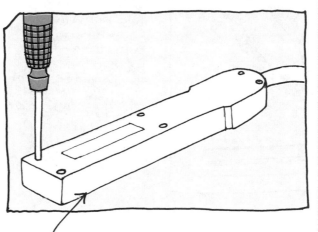

Don't forget to keep your loose screws safe!

3. EMPTY IT

Remove all of the metal connectors and cut all the wires inside.

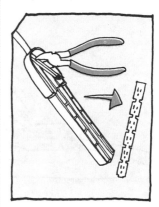

4. A LITTLE OFF THE TOP

Using the diagonal cutters and starting at the "bottom" socket (the one farthest from the power strip cord), trim away the plastic from the inside of every other (or so) socket. (Some sockets will be used to plug in the perches from Step 1 and need the structural support; other sockets will be feed holes, so clipping the plastic allows the seeds to come right up against the socket opening.)

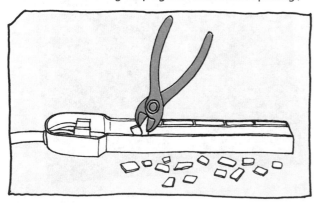

5. EPOXY

Apply epoxy to the flat part of the plugs (from Step 1) in between the three prongs. After applying the epoxy, plug the perches, one at a time, into the sockets (that haven't been stripped) and position the power strip so the two pieces lay flush. Allow to dry overnight in a well-ventilated area.

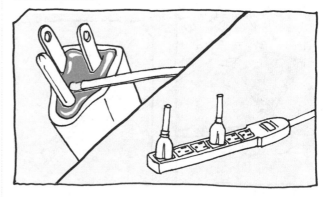

>>>

6. SWITCH OFF

Remove the switch from the power strip by squeezing the tabs and pulling it out through the front.

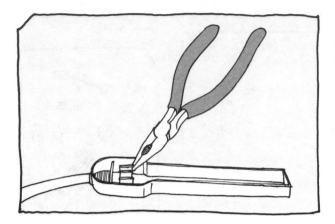

Note: If the switch is held in place by some other method that does not allow it to easily pop out and/or back in, leave it in place and drill a 3/8" hole into the back of the power strip for inserting the birdseed!

7. CLOSE IT UP

With the exception of the switch, reassemble the case of the power strip and make certain that the power cord is securely attached (you will be hanging the bird feeder from this cord).

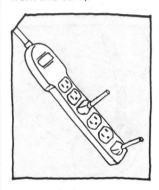

8. FILL

Use a funnel to fill the case with birdseed through the hole left by the switch. Leave enough room to put the switch back in.

9. HANG

Pop the switch back in. Loop the end of the cord over and use a zip tie to secure the cord to itself. Hang the bird feeder over a tree branch, pole, or string and wait for the birds to flock!

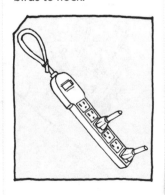

#59 Cat (Grass) and Mouse

LEVEL 1: Novice | **TECH TRASH:** Puck Mouse

Many otherwise happy mornings I have awoken to find a nice little gift carefully placed at my bedroom door by my sly little handsome killer. My killer's name is Misha, and he must really love me judging by the number of dead mice he brings home.

As I start my mornings disposing of these poor mice, I often think how I can properly show Misha my gratitude. I think I finally have a plan. I'm going to leave Misha a dead mouse of my own. And not just any dead mouse, but a dead mouse filled with nutritious cat grass. Just think, with such a delicious snack to be found inside the house, he doesn't need to spend the evenings outdoors hunting mice. I'm sure that after spending a couple of nights inside, he will find all kinds of new and special ways to repay my generosity.

> ## MATERIALS
> - A puck mouse (or similar)
> - Mini screwdriver set
> - Marker
> - Scrap of wood
> - Power drill with a ¼" plastic bit
> - Potting soil
> - Cat grass seed

Mr. Resistor Man Says:
In the absence of sunlight, red and blue LEDs can be used to grow plants.

MAKE IT

1. GUT THE MOUSE

Using a thin screwdriver, pry the dome cover from the mouse base. Remove the innards so you are left with an empty half-shell.

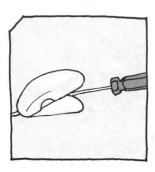

2. MARK IT UP

Turn the dish upside down (or rather put it in the original position were it a working mouse) and mark three to four spots for drilling so that the holes will be slightly elevated off the table.

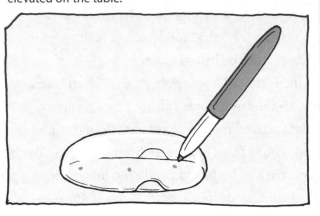

3. DRILL

Place the mouse (still dish facing down) onto a piece of scrap wood or worktable. Drill holes through the marked spots.

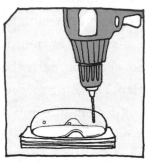

4. WASH

Use running water to wash the plastic and wood dust out of the plastic mouse shell.

5. SOW YOUR SEEDS

Fill the inside of the mouse with potting soil. Spread the seeds out over the surface of the potting soil then cover them with ¼" of potting soil.

6. GROWING INSTRUCTIONS

Place your mouse in a windowsill where it will receive direct sunlight. Water twice daily until the grass starts to sprout. Once sprouted, water as necessary when the soil starts to dry out.

7. THE GIFT THAT KEEPS GIVING

Be your cat's new best friend all over again.

Purr...fect!

#60 CAT Tube

LEVEL 4: Expert Squad | **TECH TRASH**: Old Monitor

Be they searching for porn or cats, people surfing the Internet are only obsessed with one thing—you guessed it, image quality. As the resolution of LCD monitors improves, those old cathode-ray tube (CRT) behemoths are becoming relics of the past. It's probably for the best, since those old CRT devices are some of the most toxic devices ever mass-manufactured and wantonly distributed to consumers.

But let's get back to our discussion. I'm going to assume that you, dear reader, spend most of your time looking at pictures of cats doing cute things. And just because your CRT monitor kicked the bucket, it doesn't mean you have to stop using your monitor to watch cats doing cute things. The monitor frame is the perfect size for a comfy little cat bed. Imagine the resolution your old monitor will have once you put your cat inside—it's just like the Internet—in *real* life!

MATERIALS

- Old monitor
- Screwdriver set
- Protective eyewear
- Thick rubber gloves
- Diagonal cutters
- Large cardboard box
- Hot-glue gun
- Small pillow
- Blanket

⚡ SAFETY FIRST

>> This project involves dangerous and potentially lethal voltages and implosion risk. Before proceeding, please fully read the Safety section, page 42, regarding CRT monitors.

MAKE IT

1. REMOVE THE BASE

Unscrew and remove any base attached to the monitor.

2. FLIP OUT

Flip the monitor over so the screen is facing down.

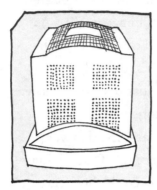

Handle the monitor very gently.

3. SAFETY FIRST!

Remove any metal jewelry from your hands and wrists. Put on your rubber gloves and tuck in or roll up any loose, hanging clothing. Lastly, put on your shatterproof safety goggles.

4. REMOVE THE CASING

Remove the large plastic casing that holds the contents of the monitor. Put aside all of the screws for later reassembly.

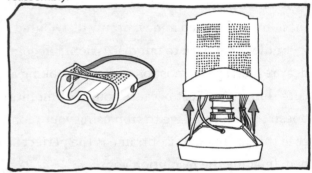

5. LOCATE THE CRT

The CRT should look like a large, upside-down vase with a suction cup over the narrow end.

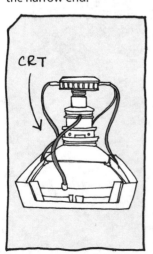

6. DISCONNECT THE CRT

Carefully figure out how the CRT is secured to the large circuit board inside the monitor. Try to unplug wires between the tube and the circuit board whenever possible as opposed to cutting them. If you do cut the wires, be sure that your diagonal cutters have an insulated rubber handle. If possible, work with your left hand behind your back to minimize the chance of a lethal shock.

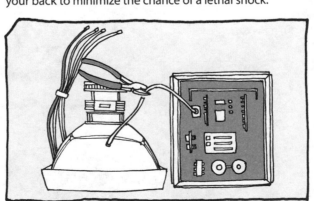

SAFETY FIRST

>> Be extra-careful not to hit or lean against the CRT monitor. The CRT is a glass vacuum with thousands of pounds of force stored inside and they have been known to implode violently.

7. SEPARATION

Separate the circuit board from the plastic monitor casing. Don't forget to discharge the capacitors if you haven't done so already.

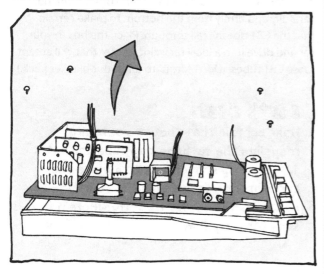

8. UNFASTEN THE CRT

Use a Phillips-head screwdriver to carefully unfasten the screws holding the CRT to the plastic monitor frame.

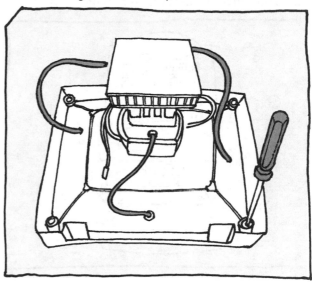

9. REMOVE THE CRT

Handling the CRT is perhaps the most dangerous thing you will do in this book. With that said, lift the CRT from the monitor case by securely grabbing the front screen with both hands. Place the CRT aside somewhere safe and out of the way. If you have a thick cardboard box handy, set it inside.

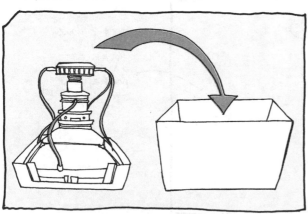

⚡ SAFETY FIRST

>> **Do not pick up the CRT by the neck or by the securing straps that may be wrapped around the tube. The straps can break easily—and, need I remind you, don't drop it!**

10. ODDS AND ENDS

Clean out any extra parts that may be left inside the monitor. Also, clean off dirt and grime from inside the case with a paper towel or rag. Collect all loose buttons and knobs and set them aside so they can be glued back on later.

11. REASSEMBLE

You should now be left with just an unassembled empty plastic monitor shell. Well, what are you waiting for? Reassemble it. Don't forget to add finishing touches, such as gluing knobs and buttons back on.

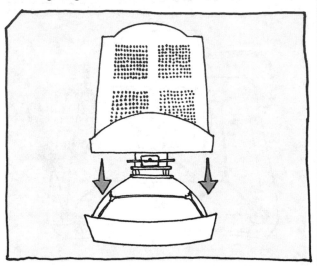

12. RECYCLE

If you haven't already done so, place your monitor, screen facing down, in a sturdy cardboard box. Wrap a blanket loosely around the CRT to cushion it in place and help in protecting you in case of an accident. Close the box. Once the CRT is secure inside the box, lift the box carefully and firmly from the bottom to make certain that the CRT doesn't fall through. Place the box in your car and drive it to a local recycling center that will accept loose CRT tubes. (Don't forget to take your blanket back.)

Earth First!

Make certain that they are actually recycling the cathode-ray tube and not just sending it overseas to pollute the Earth in some foreign landfill by asking the people at the center exactly where it is sent for recycling.

13. MAKE THIS HOUSE A HOME

Place a pillow inside the monitor casing for your cat to lie on. Decorate it as you think would be purrrfect. (You know your cat better than I—customization could mean anything from a new paint job to a disco ball. It's up to you and your cat.)

14. HOUSEWARMING

Now warm your cat up to the idea of his or her new home.

Add pillows, an old T-shirt—anything to make your furry friend feel cozy.

#61 Flat-Screen Ant Farm

LEVEL 3: Expert | **TECH TRASH:** Flat-Screen Monitor

You stared at your old flat-screen monitor for many long hours and your monitor always stared back at you lovingly with a cool blue glow . . . that was, until it died. You have since laid your good friend to rest on a shelf in the back of your garage. Sometimes you reminisce about it and wish there was a way you could once again stare into its 15-inch screen for hours on end, like you did in the good ol' days. So, what if I told you there was a way to bring it back? Come in close and let me tell you! Dust off that broken monitor and turn it into a home for our favorite industrious little friends: an ant farm, built for hours of amusement. It'll be like the Discovery channel meets DIY network meets reality TV. Look at them dig! Look at them tunnel! Look at them drink tiny sips from a drop of water! Oh, the fun.

MATERIALS

- Flat-screen monitor
- Screwdriver set
- Nitrile gloves
- Piece of acrylic to fit inside the frame
- Thin book (3/4"-wide binding)
- Epoxy
- Aquarium glue
- Clear photo box frame 2" deep and the size of your screen
- Power drill and 1/2" drill bit for plastics
- Scissors
- Insect screening
- Sandbox sand
- Funnel
- Ants
- Gaffer's tape (or other residue-free tape)

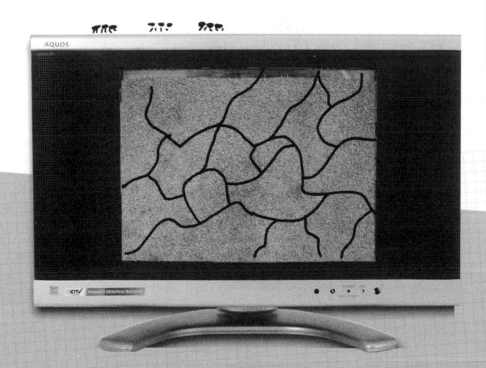

MAKE IT

1. OPEN UP

Unscrew and carefully pry open the television casing to find the main circuit board.

2. GUT IT

Remove the circuitry and screen assembly.

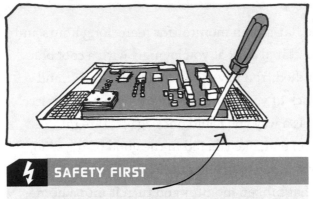

⚡ **SAFETY FIRST**

>> **Monitor circuitry contains high-voltage capacitors that can kill you. Avoid handling them if possible; if handling is necessary, see page 42 for safety.**

3. PREPARE SCREEN

To set the depth of the ant farm at 1", place a piece of acrylic on top of a ¾"-thick book. Apply epoxy around the edges of the acrylic and then gently lower a 2"-deep photo box frame around the acrylic. Let the two dry together.

4. SEAL

Read and follow the safety precautions on the aquarium glue packaging, and make sure you're in a well-ventilated area. Squeeze glue around the edges of the backer to seal it completely. (This will prevent the ants from slipping through the cracks.)

5. EPOXY AGAIN

Epoxy the ant farm into the front of your television frame.

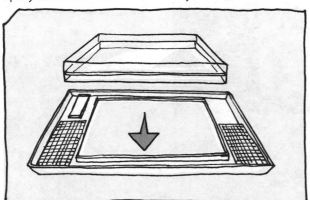

6. DRILL

Drill a hole in the top (or side near the top) of the frame that will later serve as both an air hole and access hole for occasionally feeding the ants.

In this example, there was a large hole in the back of the television that made the top back an easy access point for feeding the ants. No additional modifications to the TV frame were needed.

7. MAKE A COVER

Cut a small 1½" by 1½" rectangle from the insect screening to use as a cover. Set it aside for later.

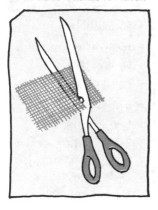

8. FILL

Line up the funnel at the opening. Pour sand into the case, filling it three quarters of the way. Give it a gentle shake back and forth so the sand settles.

9. ANTS

Line up the funnel again to the opening. Place the ants inside the funnel and cover the top to encourage them to travel down through the funnel.

10. COVER IT

Use gaffer's tape to secure the screen cover over the hole and keep the ants from escaping.

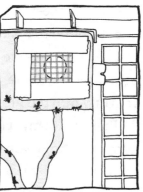

11. REASSEMBLE

Reassemble the TV frame and admire the start of a wonderful colony. To encourage tunneling near the surface of the frame, cover the bottom part of the ant farm with black paper for a week or two.

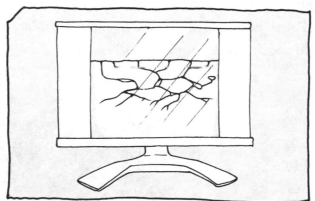

Note: Due to government regulations around invasive species, you cannot easily buy queen ants, and without one your colony will die within a few months. However, you can go outside and dig up an ant colony and find a queen ant. This will extend the life of your colony indefinitely. If you do this, be careful to avoid red stinging ants.

Proper care and Feeding

Give the ants about 1 to 3 drops of water (not enough to collapse the tunnels) once every couple of days. Also, ants love to eat honey. However, don't get carried away: You only need to give the ants 1 drop every few days. Note: Do not mix different ant colonies together in the same container. They will fight to the death, and this is not a very humane thing to do.

#62 iMac Terrarium

LEVEL 3 : Expert | **TECH TRASH**: iMac

I know the Macquarium is a classic, but it's a little boring and leaky, to be honest. It would be irresponsible of me to aid you in building a contraption to flood your living room. Besides, where on Earth are you still going to find an Apple SE? So, put away your rubber wading boots, because we're going to stop treading soggy grounds and march valiantly into the future to make an iMac Terrarium.

There are many remarkable features that set apart the iMac Terrarium from earlier Mac-based animal containment units. For starters, it won't leak. It can also hold an array of different little creatures that are more fun to engage with than a goldfish. Most importantly, it doesn't use an ancient form of Apple computing that now sells online for roughly a million dollars solely because people like making leaky aquariums. For all of these reasons and many more, iMac Terrariums are the recycled pet containment units of the future.

MATERIALS

- An old iMac
- Screwdriver
- Wire cutters
- Cardboard box
- Hot-glue gun
- Epoxy
- 3-mm plastic sheeting
- 9" by 11.75" sheet of 1/8" acrylic
- Empty gallon water container
- Pet bedding
- Pet accessories
- A pet

Mr. Resistor Man Says:
The iMac was the first major desktop computer to be manufactured in an array of colors. This changed the aesthetics of the entire consumer electronics industry.

MAKE IT

1. REMOVE THE BOTTOM

Begin disassembly of your iMac by flipping it so the screen is facing down. Remove the screw in the bottom of the computer and carefully pull off the two parts making up the bottom frame.

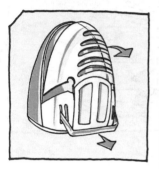

2. ELECTRONICS

Remove the electronics that are on the underside of the monitor assembly. Leave the monitor assembly intact to prevent shock hazard.

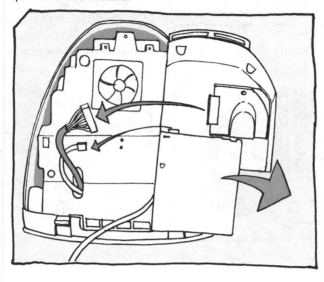

3. REMOVE THE FRONT

Gently flip the monitor on its side and tilt the screen to face you. Remove the translucent front panel. This will require gently popping it off with a screwdriver. Be careful not to crack it.

4. REMOVE THE TOP

Next, remove the top piece of colorful translucent plastic.

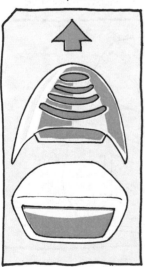

5. MONITOR TIME

Flip the monitor again so the screen is facing down. Unscrew the monitor from the piece of opaque plastic it is connected to. Unplug or clip any wires that still connect the monitor assembly to any other part of the computer. Very carefully lift the monitor by sliding your hands underneath the screen. Move it somewhere safe (a cardbord box) for the time being, and bring it for recycling as soon as possible.

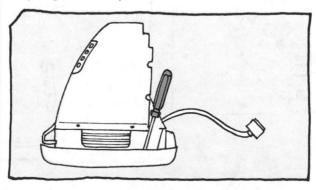

6. CD PANEL

Remove the plastic covering from the front of the CD drive.

>> >>

7. THE TOP

Start reassembly by putting the top half back together. This includes the opaque piece of plastic around the monitor screen, the colorful piece of semitranslucent plastic, and the semitranslucent sheet of plastic from the front. They should all snap together.

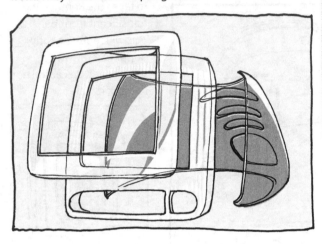

8. SCREEN

Apply hot glue around the edge of your acrylic piece and hold it, pressing it firmly, in place. The goal is to force it to follow the curve of the frame. Epoxy around the acrylic piece in place to make a stronger and more permanent seal. Place a mildly heavy object in the center of the screen and allow it to set for 24 hours. Seal in any smaller gaps still left with hot glue.

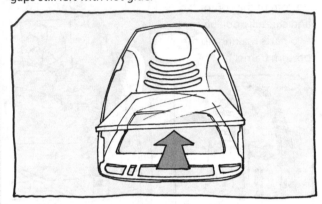

9. BOTTOM

Snap the bottom back together. This should include two plastic pieces.

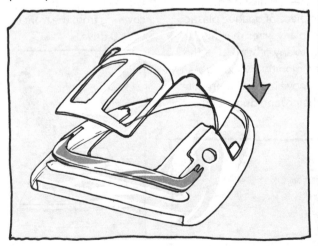

10. SEAL IT UP

Use your 3mm plastic sheeting to cover the entire bottom of the case. Completely seal around the edges of the plastic with hot glue. Cut patches of plastic from an empty water jug to cover the holes left by the drive ports and the power cord. Seal them in place with hot glue.

11. BEDDING

Line the bottom of the case with bedding appropriate for your pet.

12. DECORATE

Make it a comfortable place for your pet to live by adding a water bowl and some things to play on.

13. FINISHING TOUCHES

Snap the bottom part of the front panel back into the case. Hot-glue the CD panel in place. Seal any small holes that you may be concerned your animal may escape from, but don't forget to leave enough holes for air!

14. CLOSE IT UP

When your cage is complete, open it back up, place your little critter in its wonderful new home and snap it closed as quickly as necessary.

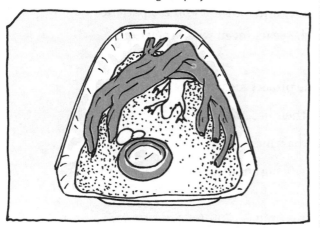

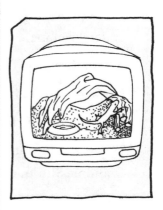

Lizards love lounging in and on their nifty new home.

WITHDRAWN
Acknowledgments

Many thanks to:

* My mom and dad for mailing me broken electronics and for their endless patience and support of all of my endeavors throughout the years (even when they don't understand me).

* Adam, Gregory, Tracy, and Eric, and for all of the project ideas.

* Mark and Susan for putting up with my stuff in their basements.

* Roman and Vida for putting up with me in their basement.

* The rest of my family . . . for something I am sure (a big shout-out to Grandma Vivian, Grandma Shirley and Pop Pop!).

* All of my friends (especially Evan, Nina, Julian, Ray, Billy, Josh, Bilal, Becky, Omar and Noah) for the good times and encouragement.

* All of those wonderful strangers on Freecycle who gave me their old broken electronics.

* Sven Travis for teaching me the value of fun and the importance of buying cheap red hats.

* James Rouvelle for a seemingly endless amount of patience in teaching me electronics.

* Everyone from Parsons DT for the inspiration and knowledge.

* Dr. Eric J. Wilhelm for sending this opportunity (and countless others) my way.

* The multi-talented Kaho Abe for technical assistance, advising, conversation and what-have-you.

* The incredible Winnie Tom for a boatload of awesome illustrations.

* And the venerable Megan Nicolay and the entire team at Workman Publishing, including Rae Ann Spitzenberger, Nancy Mace, Anne Kerman, Sophia Su, Danielle Hark, and Tom Boyce, for helping me navigate the foreign, and often confusing, land of publishing.